BRITISH PEWTER
AND BRITANNIA METAL

for pleasure and investment

BRITISH PEWTER
AND BRITANNIA METAL

*for pleasure
and investment*

CHRISTOPHER A. PEAL

JOHN GIFFORD · LONDON

First published 1971 by
John Gifford Ltd,
125 Charing Cross Road,
London WC2

SBN 70710172 7

Printed in Great Britain
by Butler & Tanner Ltd
Frome and London

Dedications

From me—to Joy, who encouraged and shared the thrills of learning and collecting progressively together.

From us—to Welch, on whom writers have drawn so heavily;
to Massé, for laying the fire with his exhibitions, and applying the match with his pioneer book;
to Cotterell, for his charming writing, and his purposefulness, particularly for his monumental effort in compiling 'Cott. *O.P.*';
to Michaelis, for disseminating much further information over the last twenty-one years.
All of them earnest research writers, to whom all pewter lovers are indebted.

Acknowledgements

When you think that this book and perhaps two more can be bought for the price of one battered Victorian tankard, and that very clever fakes exist purporting to be worth £500 or more each, it is obvious that one should not stint on literature or on every effort to learn. When you can buy, and milk, experience, it can be worth hundreds of pounds.

Let us then pay credit to the pioneers, Massé and Cotterell, for instilling interest and the basis of knowledge. Ingleby Wood, Malcolm Bell, Hilton Price—all for their difficult tasks in pioneering. Naturally errors occur in pioneering deductions and assumptions, because insufficient numbers of examples had come to notice. Therefore old editions are best read *after* more modern writings. Credit to Michaelis, who has written eruditely since the war, and has established new facts. To Welch, for his monumental *History of the Worshipful Company of Pewterers,* on which so many authors have leaned for direct quotations.

Cotterell's *Old Pewter, Its Makers and Marks* remains the one book nearest to a standard reference by virtue of the great number of marks illustrated and the attendant information on makers. Also the text matter is of fundamental use to the novice.

I would like to acknowledge the enormous help to the knowledgeable, and to the well-established collector, of membership of

The Pewter Society. Their discussions, and aired and shared knowledge, go far beyond the limitations of my book, and my own knowledge is due very largely to this Society and its members.

I must express thanks to Messrs B. T. Batsford Ltd for the arrangement to reproduce a small selection of Cotterell's hand-drawn marks, which are amongst those most likely to be met. To the Board of Trade, Department of Weights and Measures, for permission to reproduce the complete 1950 list of verification stamps identifying counties and boroughs by the numbers. To *The Connoisseur* for permission to reproduce drawings of spoon types from articles by me. My warm thanks to the publishers for letting me write this book in my own way, without interference, but with very good co-operation. All illustrations are from my own collection except where credit is shown.

Photography (great majority of) by Tom Wilson, of Robert Ashley and Partners. Drawings by J. D. Shea. Cover design by N.P.B. Ltd.

Contents

Dedications   v

Acknowledgements   vi

List of Full Colour Plates   xi

List of Black and White Figures   xii

PART I

1. Introduction   3

2. Acquiring and Collecting   14

3. Cleaning and Repairing   23

4. Marks   36
   Selection of some of the more frequent   42
   Touch marks   56
   Secondary marks   60
   Official stamp numbers   70

PART II

5. Romano-British, Early Medieval—to End Sixteenth Century   79

6. 1600–1710
    1600–60   93
    1660–1710   97

7. 1710–1820   117

8. 1820 Onwards   136

9. Scottish, Welsh, Irish and Channel Isle Pewter   147
    Welsh   164
    Irish   166
    Channel Isles   168

10. Britannia Metal   171

11. Fakes and Reproductions   176

Appendix I.   Where Pewter can be seen in Museums, etc.   184

Appendix II.   Short Bibliography   186

Appendix III. Combined Glossary and Index   188

List of Full Colour Plates

1. A dresser of pewter—the ideal setting

2. The attractive appearance of pewter not cleaned too violently

3. Pewter stripped of evidence of age is garish

4. The scale on pewter is very brittle

5. A rack of fifteenth-century spoons

6. The '1605' and 'Bun lid' flagons

7. The 'Beefeater' and 'John Emes' flagons

8. Two flat-lid flagons

9. A fine dome-lid tankard

10. The eighteenth-century 'Spire' flagon—one of the most dignified types

11. Tankard of the Woollen Guild of Norwich

12. Very rare English ball baluster measures of c. 1665– c. 1720

13. Communion tokens

14. A crested tappit hen

15. Very rare early Scottish baluster measures

16. Britannia-metal mace

List of Black and White Figures

1. A Continental maker's mark — 4

2. Overcrowded display — 21

3. Magnified view of the surface after the oxide has been dissolved off — 24

4. Pocks in pewter — 26

5. Eruptions from within — 27

6. A very early plate unwisely stripped of oxide — 34

7. Sixteenth-century touch mark — 39

8. Spoon-makers' marks from two centuries — 57

9. An early dated touch mark — 58

10. Development of the form of touch marks, c. 1660 to late eighteenth century — 59

11. Types of 'hall marks' — 61

12. The Rose and Crown as a subsidiary mark — 63

13. House marks — 65

14. A branded house mark inside the base of a baluster measure — 65

15.	Verification stamps of hR, WR, AR, WIV, and GIV	67
16.	The first Imperial Measure verification mark	68
17.	A borough stamp of 1830–78	69
18.	Superb Romano-British pewter bowl	80
19.	A typical Romano-British plate	81
20.	Anglo-Saxon decoration on pewter	83
21.	Very early flagon, possibly fifteenth century, dredged from the Thames	84
22.	A most attractive early hammerhead baluster measure	85
23.	Silhouettes of the development of spoons	86
24.	An almost perfect fifteenth-century plate	87
25.	Late Tudor dish	88
26.	Early lines obvious on the body of a measure	89
27.	Magnificent Tudor Bell candlestick	90
28.	Sixteenth-century spoons	91
29.	Mid-seventeenth-century bowl-dish	94
30.	Bumpy-bottom plates of second quarter of seventeenth century	95
31.	A quart wedge, and a gill hammerhead, baluster measures	96
32.	Late seventeenth-century provincial spoons	97
33.	Seventeenth-century spoons	98
34.	Rapid development of spoon handles	100
35.	Queen Anne portrait spoon, with a recent cast from the identical mould	101
36.	Very pretty porringer	102
37.	Seven types of salt, 1590–1720	103
38.	Latter seventeenth-century candlesticks	104
39.	Broad-rim plate	105
40.	Knife marks	105
41.	Dishes of the second half of the seventeenth century	106
42.	Narrow-rim plate rim development	107

43.	A wriggled single-reed plate	107
44.	Close-up of punched decoration	108
45.	Close-up of wriggling	108
46.	Coronation souvenir plate of William and Mary	109
47.	A flat-lid tankard	110
48.	Types of dome-lid tankard handle attachment	111
49.	The scroll thumbpiece	112
50.	Hammerhead thumbpiece, and handle strut attachment	113
51.	Bud thumbpiece	114
52.	Bud body types	115
53.	Trade cards	118, 119
54.	An early eighteenth-century flagon	120
55.	York flagons	121
56.	Plain-rim plate with house mark	122
57.	Wavy-edge dish	123
58.	Lobed decoration dishes	123
59.	Late eighteenth-century bowl in situ	124
60.	Bleeding bowl	125
61.	Eighteenth-century salts	125
62.	Candlesticks	126
63.	A fine Society flagon	127
64.	Double volute baluster measures	128
65.	Ale jug	129
66.	Measures with reinforced rims	130
67.	Tulip tankards	131
68.	Range of eighteenth- and nineteenth-century tankards	132
69.	Chart of dating features	134, 135
70.	The bulbous measure	138
71.	Dating details of bulbous measures	139
72.	West Country measure	140

73. Late wriggling, and linear engraving   141

74. Beakers and footed cup   142

75. Doll's-house pewter   144

76. Mantel ornament and tobacco jar   145

77. Art Nouveau pewter   145

78. The tappit hen   148

79. Scottish Communion cup and flagon   152

80. Laver   153

81. Engraved flagon and dish   155

82. Scottish measure handle feature.   158

83. The potbelly   158

84. A thistle measure   160

85. Scottish ball and embryo shell measures   161

86. Scottish lid centring rim   162

87. Glasgow and Edinburgh measures   163

88. Possibly a Welsh flagon   165

89. Irish chalice and flagon   166

90. Irish Haystack and noggin   167

91. Irish spirit cup   168

92. Jersey and Guernsey flagons   169

93. Britannia-metal cream jug   173

94. Maker's stamp on Britannia metal   173

95. A small range of Britannia metalware   174

96. A horrible Britannia-metal teapot   174

97. Fake measure, showing close-up of the surface   177

98. Fake house marks   178

99. A fake spoon   179

100. Very fine restoration work   180

101. Reproduction pieces   181

102. A collection of fakes   182

PART I

1. Introduction

'. . . you long for simple pewter' (W. S. Gilbert)

I do not know of any 'simple' pewter. One longs for pewter perhaps for its subtlety in pure design, for in this respect there are few media to equal pewter of English manufacture. Continental work is often very much more ornate, and of inferior craftman-ship and material—but in this book we are not concerned with any pewter other than British. In fact, to clear the way at the outset, in case you should wish to exclude continental from your con-siderations and collection, there are one or two guides to enable recognition of continental work which few previous books have related.

Europe covers a large area of widely differing cultures and tastes, so it is only to be expected that there is a tremendous variety both in period and in regional design. However, differentiating from British is simplified greatly when one considers how continental pewter has come to be in this country in the first place. Migrations of refugees in troubled times travelling light would bring little, and leave proportionately few, surviving pieces.

By far the greatest quantity of continental pewter in this country was brought back by the Tommies and officers of the First World War, and at best comprises eyecatching eighteenth- and nine-teenth-century ware of Dutch, Belgian and French manufacture. It mainly consists of cylindrical measures, wavy-edge and plain-rim

3

plates, and Normandy flagons. Since then there has been a very large quantity of reproduction work for the successful titillation of tourists.

So we are reduced, broadly speaking, to Dutch and French work, although dealers from across the channel have been recovering their local products assiduously over the last few years.

Continental work is usually stouter, more leaden (and therefore disproportionately heavy) and more 'cheesy' than English. Very often plate rims are more upward-curved than their English counterparts. Twin acorns were used very largely for thumbpieces, which no English pewter bears, except Channel Isle measures (see later chapter). Lids are often heart-shaped—so are the Channel Isle measures. Makers' marks containing angels can be unequivocably considered continental. Such words as 'Fin' or 'Zinn' are as indicative of continental manufacture as any surname initial 'Z'.

Fig. 1
Typical Continental maker's mark. Dutch, c. 1700 Note the lines in the rose, and initials in the crown.

Where, as is very often the case, a Rose and Crown is struck with the rose filled in with many fine radial lines, or the crown containing initials, you can be sure the origin of the piece is nearby continental.

The same applies to pieces which have a maker's mark struck three times. There are many other less obvious indicators, but these few tips should enable you to distinguish at least 90 per cent of foreign pewter which you are likely to meet.

As we have already started to brush with pewter—what *is* pewter? In its broad sense, it is tin alloyed with lead or other grey metal. So this definition must include Britannia metal, but hitherto one need not have been classed as a pedantic purist to have segregated Britannia metal as a separate entity of ware. The latter's products are in the main easily distinguishable from pewter by design. The alloying metal is not lead but antimony, and its method of manufacture differs basically. On the other hand, in the more accepted sense, pewter is an alloy of tin and lead, and formerly copper took the place of lead in fine metal. It never intentionally contains

4

silver; traces, if any, are due to its presence in the ore, and to remove it would have been very costly; its presence serves neither harmful nor beneficial purposes in the minute proportions which occur. This point is aired because 'real silver pewter' used to be included in the well-meaning hard sales talk of antique dealers.

The standards of alloying were simply and exactly laid down in our earliest surviving records, going back to the fourteenth and fifteenth centuries. Control and observation of the regulation were well maintained, for failure to do so soon entailed fines and seizure of faulty or sub-standard products. However, evasion and enterprise developed towards the latter part of the seventeenth century, and by the mid-nineteenth century the Worshipful Company of Pewterers, who were the controlling body, no longer had effective control. In the nineteenth century manufacturers suited themselves as to what material they were using, and little or no research has taken place as to the alloying of this period.

In earlier days, the mixture was laid down as 112 lb. tin (a standard ingot) with 26 lb. of copper for flat ware, i.e. plates, dishes, etc.; and with 26 lb. of lead in place of the copper for hollow-ware, the self-supporting strength of a cylinder was appreciated. Dr Homer, a member of the Pewter Society, tells me he cannot persuade tin to accept as much adulterant as laid down, but nevertheless there are records of pewterers being in trouble for not maintaining these proportions. Specimens with copper alloy are very rare. I have only one such plate in my collection (of c. 1490) which demonstrates a quite different result from that of lead alloy —it is much harder, more resonant and apparently almost infinitely less susceptible to oxidisation. Therefore we can deduce that copper was very seldom used by the mid-sixteenth century. Lead was the sole adulterant from then until the latter eighteenth century and it would be extremely interesting, if only academically, to know the range of existing examples' proportions in the succeeding periods.

To turn to Britannia metal briefly—this alloy was first produced in the latter part of the eighteenth century, but it is now very unusual to find specimens earlier than about 1830. A separate chapter is devoted to this subject, which surely must soon be recognised as 'collectable'. It was evolved for economy, better properties in use, and capable of copying silver styles more closely. It comprises tin, 90 per cent, and antimony, 10 per cent. This produced a harder alloy, which was practical in a much thinner gauge and could be worked like the ornate styles of silver, yet still retain strength. One might say that Britannia metal is the spurned poor relation of pewter, yet this Fitz-Pewter is sometimes very charming; and in any case was the final death blow to its forbear. Cotterell, the first real authority, regrettably and with the purest intentions did a great disservice to unborn potential collectors, for in his writings he lost no opportunity to decry this medium—and so the very people who had an affinity for base metal were advised to weed out Britannia metal and be rid of it. Therefore so much has been previously disregarded, and has gone for scrap, recognised but not appreciated. Having inspired neither interest nor followers, research and publications concerning Britannia metal have been negligible in Great Britain. Nevertheless, it is interesting to read Cotterell's views on the subject, although some statements are now known to be not quite accurate. We older collectors have been under the influence of Cotterell, and few have really considered including Britannia metal in collections. Americans have been wiser—and it forms a larger and relatively earlier proportion of their pewter history—so they collect Britannia metal of American and British origin.

The same illustrations have recurred in many books and magazine articles. In this book examples are almost entirely unpublished, or little published, or in some cases have been re-photographed. More use has been made of close-ups, for intimacy of detail. Pieces are not necessarily the finest known, but are representative, and it is hoped that the illustrations will be the more appreciated since the diversity will add considerably to the collector's library and knowledge. This book is intended to be a really useful addition to the shelf of would-be serious collectors by bringing many new facts and thoughts on connoisseur periods, and also to be a vade-mecum for middle and latter period men. (There are some corrections to errors and misleading statements and impressions from earlier books.) Furthermore, it can encourage a deeper knowledge by the very many people who possess some pieces, and would like to further their information—and can introduce many more still to the pleasure and opportunities of the nineteenth century. Why turn away from Britannia metal or Art Nouveau? In twenty years' time one could rue it. It is very much hoped that this book will be of more help to the newer, and future, collectors than anything yet published and that it will fill in many facts which were not hitherto appreciated. A collector thinks not only of the past but of the future, especially when he is safeguarding his hard-earned and ill-spared capital. No-one who has collected pewter (except perhaps those who through inexperience have acquired an undue preponderance of fakes) has had any financial reason to regret his outlay—far from it. I make no secret that my own greater love is for earlier pewter, and it is more difficult to write about nineteenth- and twentieth-century pewter, and Britannia metal, because so far it has been spurned, seldom mentioned in any book, and has not had the advantage of collectors' discussions. Here are fields for research and 'top-of-the-treeism' for anyone interested. In fact I give only a basic outline of nineteenth-century ware, and expect to intrigue sufficiently a few people prepared to devote time and effort to further research.

Do not expect to find prices quoted. It is a very great disservice to both collectors and the trade to attempt to do so, and they are so erratic. It is a free market, and let it remain so. At sales sometimes two people go mad—from desire, jealousy or ignorance. I have seen examples of each in the last few months—when prices fetched were about three times those expected, and yet other prices have been lower. One can soon establish, by observation, the normal price of nineteenth-century pewter. But it is very foolish to lay down prices of rarities—however much individuals would like them. With any such rarity, and a comparatively narrow market, two or three large sales could very possibly depress prices. Stock on the market, economic conditions (both here and in America), condition of items, embellishments (such as engraving), marks, doubts about the entirety or authenticity, will all have a very strong effect on prices. If you see prices 'quoted', you can only view them with great suspicion.

So prices vary. If you go to a shop in an out-of-the-way place, you can usually expect to pay less than in a glamorous area on the American tour route—but it does not always work out logically. A dealer will sometimes price up to what he thinks he might get. From certain major salerooms one can buy lists of prices fetched, subsequent to a sale. Conversation with experienced collectors may elicit some idea of values, or one can visit specialised dealers. It is quite certain that although the author is of little consequence, publishing this book will stiffen demand and prices. So, if you are acquisitive—read fast, move fast and often, search deep and discriminate well.

Occasionally enquiries are received from embryo collectors, seeking a Society to teach them from scratch. No national Society could exist with this purpose. What are books for? By all means find contact with other kindred souls, perhaps more experienced. Most members of The Pewter Society are delighted to find learners and will give assistance—perhaps on to the stage of proposal for membership.

Let us run through a very brief history. Pewter was evolved by the Romans in Britain in the third and fourth centuries, and they made a remarkably wide range of sophisticated wares. It then disappeared until the eleventh century, in which there is record of church use. Twelfth-, thirteenth- and fourteenth-century tombs of monks, when disturbed, disclose sepulchral chalices—but without doubt there was by now considerable manufacture of widespread domestic pieces. Ordinances of 1348 establish the standard alloy, and make it clear that pewter manufacture was a recognised craft. A little more than a century later the London pewterers were granted a charter and the industry was thriving in other centres. Records relate to a wide range of products, but we cannot identify what they were, and very few examples, except spoons, survive. The records give much information concerning the sixteenth century, and although rare, there are a fair number of plates still in existence, but almost no hollow-ware. The seventeenth century was the 'Golden Age', and many examples of domestic ware of the latter half are now in collections, and also church flagons of the whole century. Competition from other media, and perhaps too many tradesmen, both caused a recession and undermined the Pewterers Company. In the eighteenth century styles became longer-lived, the Company lost control, and towards the end Britannia metal appeared. In the nineteenth century a big impetus was caused by the implementing of Imperial Measure—but only for pub tankards and measures, to which use pewter manufacture was more or less confined. Britannia metal took over domestic ware. In the first decade of the twentieth century two exhibitions were staged and the first books appeared. Awareness and collectability were created and the Society of Pewter Collectors was founded in 1918. Reproductions and fakes appeared in the '20s, and Cotterell's monumental work, *Old Pewter, Its Makers and Marks* ('Cott. O.P.'), was published in 1929. Finally, in the early '30s pubs were no longer allowed to serve beer in pewter unless asked to do so. An enormous amount of pub ware was turned out, and the last market of functional usage had gone.

Pleasure

Now let us consider the pleasure and investment aspects. To build any collection entails an enormous expenditure of concentration and effort—preferably a time and patience effort rather than buying big too soon. So many collections in other media are of numerous, widely distributed items, well known by type and market price. For instance, silver is commonplace compared with pewter. Pewter of various types and ages is scarce—very scarce—and even more so in relation to demand. It is most rewarding to one's ego to recognise and buy such items at a reasonable price. One feels a continuity of intimacy and personal ownership. Pewter is something you live with, in intimacy, not put behind glass windows, but ready always to be handled. If you have an affinity for metals, then pewter is a most satisfying medium, as many examples are to be found amongst which a desirable piece may lurk hitherto unappreciated. Almost unconsciously one is aware of its part-personal ownership, its association with the past generally, and whether it evokes Pickwick or Pepys is only a matter of degree. It looks, and feels, and is old. Pewter is a sufficiently off-beat subject to keep its adherents not too numerous, to enable the collector to be individualistic. Like all collecting, it can carry a social cachet—maybe the Joneses next door do, or do not, collect something. It results in constant awareness, seeking sources, parallels, information, museums, friends, churches, collections, shops—any is a potential source of interest. Maybe you trace out the maker, where he lived—and conduct further research on the individual and his environment. Cotterell in his *O.P.* gives ample lead in this connection—yet there is a wealth of research still to do, in order to understand more about the subject as a whole. Perhaps you like the touch and feel of pewter—to glance at its sturdy subtle lines, to see its soft glinting reflections. Probably you like the intimacy of cleaning it—not merely when bright, but removing the unsightly dark scale which so often obscures the bright metal. By handling and restoring pewter you learn more about its manufacture, discern fake, foreign or made-up pieces. Hand in hand with such participant handling come knowledge and experience. Let your collection grow in step with your knowledge and be careful not to let the former outstrip the latter—or you will regret it. But if your knowledge leads, you will take greater pleasure in finding worthwhile items for your collection and revel in the one-upmanship of having greater knowledge than the vendor.

We are all acquisitive—so go and browse in antique shops in the hope of buying. Take the opportunity to *handle* pieces. There is a rare leap of joy when you see that you are in luck. But it is not necessary to own pieces. Massé, the first writer on pewter, proclaimed he never owned a piece. Yet his knowledge, all self-taught, was wonderful pioneering. He put it to good use by staging two fine exhibitions at Clifford's Inn. With your growing knowledge you can spot things of an age hitherto unsuspected, unusual items in the alloy, something new—or fake. You could be able to help museums in the identification or dating of pieces. One of the joys of pewter is that there is so much more to learn—paradoxically particularly in the nineteenth century—and there are more pieces of that period to be found than of previous centuries. Its wares were too commonplace in the 1930s, '40s, '50s and '60s—but now they are thinner. Hurry, and record and note everything so far unwritten.

There is a great excitement in tracing makers' marks—perhaps to find they are unrecorded, or perhaps to add more to existing knowledge.

In the home it is not easy to display pewter in quantity with modern stark décor. Obviously a dresser, shelf or mantelpiece is excellent, and all are suitable positions. Be prepared for growth, and its pleasure. Be prepared, too, that you may start off with one or two pieces—and a little reading—increased enthusiasm—a surge of searching—some luck—and you become an ardent devotee. For a tolerant companion in the house—something basic and homely, with its soft gleam broken up by a myriad of dents and irregularities, yet not fragile—what medium is more peaceful?

Investment

There have always been collectors, even in prehistoric times. Their emphases vary over research, preservation, love, possession, investment. The first abhors the last, and this thrives solely on the demand created by the others. Yet few people make a collection of any standing without, amongst other attributes, seeking the comfortable feeling that the money is well and safely spent, and easy to realise. Some in the past will have been skilful enough, after a short initial period, to have increased their collections at little extra cost, by selling their earlier, more commonplace items, and buying true finds at advantageous prices. Now, so many books are written on antiques (and are soon out of print): also by magazines, press, television, lectures and exhibitions knowledge is disseminated, so would-be collectors and the trade are more widely educated and to a higher standard. Prices are higher and more uniform. This has brought more out onto the market and circulation by recognition. The holder perhaps sees his collection change from intimate brothers and sisters to haughty glamour pieces, almost remote from the touch (in other fields even to be banished to the bank).

When your search has at last led you to a real find, factors on confrontation with a piece are rarity, demand, condition, authenticity, price. Be bold from knowledge, not rash with optimism. How often one finds a collector who has impatiently rushed his fences—spent more money than sense—and has bought a fake, made-up piece or reproduction. One sees it so often, and I have done it myself, when more keen than knowledgeable. So often desire is greater than discretion. To collect for pleasure and investment, you must balance desire with knowledge, discrimination and money available. It is this latter factor being so out of gear with experience which has led over-confident lone-wolf collectors to waste away resources on fakes, or half-fakes. Why is it that an ignoramus of a subject so often seems to think that the piece he buys is *the* find of the year. This is the hazard of brashly stepping out of his sphere. It does not mean to say that with taste and background you cannot make a killing. I did not really know what my first piece of good pewter was, but fortunately had enough experience of oxide and general form to have found an exceptionally rare type of seventeenth-century flagon. Later I bought a fake glamorous candlestick, a reproduction tappit hen, and later still, a faked Horned Headdress spoon—the first and the last of these from high-class well-known shops. The fact that neither shop should have possessed, offered or purveyed such goods—and very fully priced—is beside the point. The responsibility is the purchaser's. He must be the judge—and judges are wise, experienced and observant men.

Where is the profit? We have seen that the dissemination and acquisition of information increase yearly. A little more spare cash jingles in many prudent pockets. The car yearns for a spin in the country—an antique shop comes into sight. . . . Knowledge, transport, emergence, distribution, competition, desire and cash all increase the demand on a limited 'stock'—a demand which can only be met by a redistribution of existing items, amongst more seekers with longer pockets—the values are greatly enhanced by the perpetual auction in being. Everyone bids higher, pays more.

Let the collector to whom taxless appreciation is the goal also engender and extract all the fun of learning and seeking. One who acquires and hoards without putting back into the subject (and arrests its study by frustrating the eager would-be scholars) is utterly abhorred by the connoisseur, for he has pushed up the price and removed the goods from view. The true profit in collecting is pleasure—for yourself and others.

On appreciation, two points. No one piece has yet broken the £1,000 mark, at which Capital Gains Tax is applicable—but the day is not far ahead. Lest you are about to slam the book, decorative and interesting pieces of nineteenth-century ware can still be bought for a very few pounds. And do not forget to review the value of your collection for house-insurance purposes, say every two years.

Taken all round, prices have increased, roughly speaking, I suppose, about threefold over the last fifteen years or so, and this is particularly noticeable of nineteenth-century tankards and measures. It is only reasonable to expect prices to go higher at a similar rate, barring a serious national or international financial setback.

Finally, do not forget that any piece has two prices—to buy and to sell—what you buy at and what the trade buys at. Do not grudge the trader his turn. He has driven many many miles, advertised, paid his rent and rates, and displayed his wares, for you to see at your leisure. Furthermore, this is his livelihood—so you must expect to wait two or three years before you can sell back at your purchase price.

Maybe you are impatient to get to grips with detail—by all means do some leap-frogging. Turn to the chapters you want—but do come back! You may have some possessions, and only want to know how to clean them. May I invite you to seek more detail as to the date and background of your mugs or plates. For this reason there are some early action chapters—e.g. 'Cleaning', 'Marks', etc. However, inspecting the background in all connections is a shortcut to many years of expensive and hard-won experience, and perhaps lost opportunities. It may be asked by new and future collectors, why waste time in delving into facts which are purely archaeological and archivistic? They are NOT. You never know when you may meet something. Not long ago I found a sixteenth-century salt, in poor shape, supporting a stone ball, acting as a door stopper in an antique shop. You must know the history and background in order to appreciate, put in perspective, and know what you are associating with. Since starting to write this book I have gained notably a Tudor dish, a broad-rim dish, and two fine dome-lid tankards (and not from major sale rooms, or specialist dealers). Never despair, you too will see pieces —but they come when and how you least expect.

The format of this book is a little unusual, with its early chapters on practical aspects, such as Collecting, Cleaning, and Marks. Furthermore, instead of dealing with each type in turn, which has much to commend it, progress is made through the periods in order to keep styling of pieces of an age in close company, and also for those who wish to consider an era as an entity, to view the scene with its examples in relationship. The products of the eighteenth and nineteenth centuries fall fairly neatly into their centuries and these are the periods whose specimens one is most likely to encounter. However, one or two specific items are taken over their whole run where they overlap a period by only little. Fakes and repros. must be dealt with separately, of course, and not in their purported period. Britannia metal, too, must be taken as a group apart, both for recognition and distinction, and because, for the greater part, its products are so dissimilar to pewter. People want to know where they can see pewter in any quantity. There are some specialist shops in London, and a few provincially. Museum displays are few and are noted in Appendix I, but sometimes a display is mounted for a short period. Museums known to have collections not on display are noted, and prior contact is advised for those wishing to see them.

There is a short bibliography in Appendix II. Some books may be out of print or hard to get, but public libraries will always get them in—but possibly for reference only.

Appendix III is a combined glossary, index and illustration reference. However, technical manufacturing words have seldom been introduced into the text, and in fact technical terms have been avoided where unnecessary. For instance, the pedantic distinction between a tankard (lidded) and tavern pot (lidless) has been very willingly suppressed. I have yet to hear one of these pedants ask for a 'tavern pot of bitter' at the bar.

Finally, there are several blank pages for your own notes. You can increase the value of the book, at least to yourself and probably others, by recording your own notes and observations, as, out and about, you encounter items of interest, perhaps to be written up at home.

2. Acquiring and Collecting

In this day of vast takeovers and world-wideness, many people are so eager to plunge head into sand back to the aura of individualism and personal possession. They seek links with past domestic use, with long dead but 'live' people, and with historical times. Man is acquisitive, and competitive, so he acquires as a competition with other people, for one-upmanship over the Joneses next door. What better than to collect something which is rare and not well known; expressing individuality by unusual knowledge; acquiring pieces which are scarce but only recognised by few; old and battered yet which one can restore and repair; restful, homely and attractive in display; and where one can start sensibly with modest expenditure—or can open up to pieces worth a car apiece. These qualifications are met by only a few classifications of 'collectables'—and surely pewter answers all of them.

We mentioned 'rare'. Why *is* pewter now so scarce, which in the seventeenth and eighteenth centuries was so commonplace? How has it come about that early pewter is in such short supply? There are many reasons beyond normal wastage—or perhaps it is more true to say that normal wastage encompasses more reasons than other media.

Pewter is soft, so it soon shows signs of wear with its many dents. And although flexible, it is liable to tear, e.g. where the handle joins the drum. Its softness renders the rim of a plate or dish liable to part company from the well. Its very tenderness makes it more susceptible to wear than most other media. Surely it was easy to repair or replace parts? Yes—as a matter of fact, too easy. The Pewterers Company in the fifteenth, sixteenth and seventeenth centuries combined the two functions of a Masters Federation and a Trade Union, and all under a monopoly! They said 'No repairing'—so damaged pieces could only be traded in as scrap metal. Jobs for the boys indeed. So already we see that pieces were liable to damage and were then only good as scrap.

Styles changed surprisingly quickly. It is a fallacy to think of fashion and style as static—certainly in the seventeenth century. In the eighteenth and nineteenth centuries in pewter they were very slow moving—but there are usually details which aid dating. Earlier, the whole conception of styles and their imposition was fast moving. Items—say salts or spoons—were quite inexpensive and so, when out of fashion, would probably readily have gone into the melting pot. Those items which were not disposed of, in the course of a year or two, would lose their brilliance and become coated with a gradually thickening scale of dark unattractive appearance. Spring cleaning, or the seventeenth-century equivalent of a jumble sale, would cause them to be discarded.

The very commonplaceness of the medium would militate against interest, or foreseeing our avidity now. Think how even small pieces of silver, a precious metal, were kept—not so in pewter. Many of these reasons explain why silver, unlike pewter, abounds in such profusion. As well as being out of fashion, or out of sorts in texture, other media competed, threatened and took over its uses—pottery, porcelain, bronze, brass, copper, Britannia metal, plated goods (and silver itself—with rising incomes). Is that enough? It is more than we like, but not the whole sad story of the wastage.

Many of us remember the urge and patriotic frenzy with which householders gave up all sorts of metal in the last war, from aluminium saucepans for Spitfires, to gold sovereigns. With the vast industrial calls for tin the scrap-metal merchants have been keen to buy all the pewter and Britannia metal possible at such attractive prices. What desirable damaged pewter must have passed through their hands—and early Britannia metal too. Just the same happened in 1914–18 for war purposes, and thereafter. And it is not generally realised that the situation was far worse in the era of the Napoleonic wars—in the 1700s metal was desperately short. So much for the scarcity of the pieces.

Now look at collecting history. Following Massé's two exhibitions in Clifford's Inn and his book, which all came in the first decade of this century, there was an upsurge of interest in collecting. In 1918 the Society of Pewter Collectors was founded. (This is now the Pewter Society, appreciating that the 'Collectors' tag may inhibit really keen non-possessing individuals.) This fact, and magazine articles, stimulated the search. Cotterell produced his magnificent book in 1929 with drawings of over 1700 makers' marks, and invaluable text on various general technical aspects of collecting. This was subscribed for by almost 200 collectors and institutions—at home and abroad. Since then the Second World War introduced vast numbers of American and Canadian troops especially to the delights, and removable (paid for, of course) décor of our pubs. Similar tankards were sought and bought in shops. Subsequently we have run into the affluent age in which there is a little spare cash often in the pockets of all classes, and many use it to collect something or acquire varied bits and pieces. Almost everyone is mobile according to their own volition, and can undertake, not so much a pub crawl, but a pub tankard crawl, and potter from shop to shop. Education is widespread—so is availability of taste. Literature abounds; magazine articles often deal with specialist technical subjects. All this applies not only to the United Kingdom but Scandinavia, the Low Countries, Italy, America, Canada, Australia, etc. An enormous amount of nineteenth-century pewter has threatened to sink the cargo ships to the States. Now comes the question whose answer is liable to be so misleading. 'What am I likely to come across?' One is tempted to be stuffy, and say, 'Go out and note what you see, that is what to expect.' It is not as simple as that, for already nineteenth-century pewter in good condition is of considerable value, and ascending in inverse ratio to supplies appearing.

You will come across a lot of nineteenth-century pub tankards, and if you study the chapter in this book on 'Marks' you will be ahead of everyone who has not taken such trouble, for much information is published here for the first time. Other chapters will help you to spot the once in a lifetime find. Diligence and observation result in finds of far more favourable frequency. These pub tankards and measures are tough, solid and attractive. Little is yet known for certain about the origin of styling and this is

Plate 1 Plenty of space gives an appreciation of each piece, and a feeling of restfulness.

Plate 2 A close-up view of the most attractive condition in which to display pewter – bright, but not stripped.

another field for research. There is always the possibility of finding a somewhat similar tankard with perhaps the fillet in a slightly different position—and you may be in a world 150 years earlier, and you have made a kill.

You will find plain-rim plates and dishes of the eighteenth century, and hot-water plates and plates of the nineteenth century. Many marks or subsidiary marks are, you will find, not recorded in Cotterell's *Old Pewter, Its Makers and Marks* ('Cott. *O.P.*'). Always take some paper on which to take rubbings of marks with you— and it is very helpful to take a camera to record pieces which for various reasons you cannot acquire. You will find a great variety of more or less pleasant nineteenth-century pewter. I cannot become pleased by bed pans, even with flowers. Syringes for dealing with the intimate troubles of animals are out of place in my, and I hope in any of my hosts', dining rooms. But tankards, measures, snuff boxes, mantel ornaments, casters (for sugar, pepper or pounce), salts, perhaps bowls, are to be had.

You will come across rowing trophies of the 1880s and '90s, and Art Nouveau pewter; Britannia metal—mostly late, gross and over-decorated pieces—but however much we dislike their design, these are tomorrow's collectors' pieces. In Britannia metal, examples will range over a bigger field—teapots, coffee pots, church plate (flagons, cups), hot-water plates, tobacco jars. Get to know the makers and their dates and use your knowledge and taste. Britannia metal was also used as a base for plating—either beware or belove, according to your taste, pocket, and will for self-expression. But do try to select for style and date, rather than mere existence. One should add, and we will talk at greater length in a later chapter, that you will find reproductions—certainly I hope you will recognise them. If you wish, collect them—but I do not advise it, as they will not appreciate in value for a long time; and you often have to pay more than their true value, for shops frequently have not realised their lack of the age they simulate. You could make a fine collection of fakes—but I fear you would have to pay much more than their worth, for it would seem that some unscrupulous dealers are snapping them up for export, and the prices commanded indicate competition and lucrative trade. Collect from strength of knowledge, not strength of pocket—read, see, observe and handle all you can. Herein lies the thrill of collecting.

It is often assumed that connoisseurs are collectors. What about the Egyptologists, archaeologists, art critics? Probably they will not possess a single example of their affinity. Neither need you with pewter—but once one has handled, nursed and cared for it, there will be a subconscious longing to own. Flirt with it. Flirt outrageously. But you *can* be a connoisseur, and researcher, without possessing. What a collection you could make for yourself of photographs—including intelligent use of ultra close-ups. You could even learn to *make* your own collection—from photographs, drawings and knowledge; perhaps in miniature. Learn how to repair—here is a satisfying spare-time job which could pay for collecting and with all the opportunities to learn from handling—and possibly a prior option on pieces!

Where to seek? Most antique shops seem to specialise within the knowledge of the proprietor—but occasionally, say, a glass man will have bought a lot which may have 'mucky metal' in it. I personally prefer the few informal 'dirty' shops in which one visualises unrecognised treasures. It is worth looking at bric-à-brac shops and charity shops—it is a question of your available time, persistence and patience. Auctions, large and publicised, will attract specialist dealers and collectors. Country sales will be infested by shopkeepers and runners, not to mention possible collectors, and the myriads of saleroom haunters. Do not forget that a sale at a house of 'stature', or whose occupants were well known, may for naïve reasons entice buyers to think that everything is first class—overlooking that the best may be held back and that the sale may be packed up with a lot of other virtual trash. Read your sale catalogue carefully—you don't want to buy a flat-lid 'Charles II Style'. Inspect the pieces most critically and set your price before the bidding. Auction frenzy is infectious and dangerous to your pocket. At the major salerooms the prime pieces (and lots of others) are to be had, but you are unlikely to get bargains. I have seen several embryo collectors go out and pay high prices, but backed by no critical faculty—and in almost every case they buy something they did not know about—a new thumbpiece has been applied—or worse. There will be other collectors, shops buying for their own stock, even professionals buying for their own private collections, and—even more 'enigmatic'—the dealers with an export market to cater for. Wherever you go, you must be your own judge. A story to amplify—in a specialised shop twenty years ago, I decided to test their ethics. I said I only wanted English pewter, and asked ordinary elementary questions, to establish that I did not know much—and then handled a piece I could see was foreign. Reiterating that I only wanted English, I asked the price . . . never a bleep to give me advice. This is NOT true at all—but it serves to show that you must take your own objective decision which you can only do from knowledge.

Massé, Hilton Price (for spoons) and Cotterell all blazed a trail. Michaelis has edited Massé up to date—but if you read the original, and the others, although 90 per cent perhaps is as accurate as the day it was written, you must allow for pioneering misconceptions—and you do not know where the odd 10 per cent lies. I have no doubt that despite the confidence in which this book is written, to the best of my experience and knowledge, now or in years to come there will be some facts or statements which will create a snort from more complete knowledge. Statements are made apparently authoritatively and it is understood that this is always with the proviso 'to the best of present-day deduction and knowledge'.

This question of building your own confidence extends further. Cotterell most regrettably lost no opportunity to decry Britannia metal—to tell you how to differentiate, so that you could 'lose' it. Since then, apart from in America, where it forms a large proportion of their shorter pewter history, it has been 'beyond the pale' for respectable collectors. And so large quantities of delightful pieces have been scrapped. Do not be bound by Cotterell (*O.P.* was written in 1929 and the situation has changed). Collect what you like, and add to your knowledge by research. The earlier Britannia metal is extremely pleasant—comparable in form to silver—and to my mind, should be sought eagerly but with discrimination.

It is dangerous to let it be known what you seek, specifically. I do not know if fakes are being made today. I hope not, but expect so. In the '30s the unsuspecting collectors let it be known that they wanted—say, a half-gill hammerhead to make up a range. Surprisingly, within very few weeks one would be purveyed, with glamorous house marks, very 'old', battered, and repaired; technically made with complete knowledge of experts. It may happen today, tomorrow. Collectors in all fields appear to be very hidebound and timid, and are inclined to buy only examples of what is 'in the book' and illustrated. Several times I have come across an obviously old but unpublished type, which other collectors had seen and eschewed from temerity. Thus 'new' types can occasionally be discovered at a fraction of the price of pieces nearly comparable.

Several enquirers expect to join a Society to teach them from scratch. At present there is only one—the Pewter Society (formerly the Pewter Collectors Society)—but this wishes to remain confined to 40 members for wieldiness and for intimacy in advanced discussion. It is therefore esoteric. You do not join the M.C.C. to learn to play cricket. In fact, you probably have a playing-in standard at your local tennis club. But those who have been through the mill, and will probably be able to add further benefit of experience in the future, may be invited to join the Society. It is purely amateur. It will try to give assistance to individuals, as well as museums and the Church, and will also try to arrange for talks to Antique Societies, etc. There is also a large club in America —the Pewter Collectors Club of America—and a small group in Holland.

It is regrettable that most collections are a number of possessions with no known history or provenance, which, in the sum of knowledge, would be of inestimable value. So many makers' marks only bear initials—and we do not know who or where the maker was. So many marks with full name do not divulge the town of origin. We know little of regional types and so very little about the pewter of Wales. Try to find out true history and provenance of pieces, but beware of fascinating and entertaining antique dealers' lore.

It is rather optimistic to decide to collect one period, or one type. One has to play it *ad hoc*. Of church pieces, e.g. flagons, unfortunately theft has been rife. No piece of church property may be sold without the Bishop's faculty—and the sum raised must be invested and only the interest used. But of the many flagons (church or possibly domestic) in collections, remarkably few are still accompanied by faculties. We tend to be very insular, and most people collect only British or English pewter. A few years ago continental pewter was available at very low prices. Since then it has been most eagerly sought by dealers from Scandinavia and Holland.

With any antiques it always pays to get the best condition pieces, and beware of made-up items. There is a temptation in early days to add numerically, and to accept perhaps a new lid on an otherwise entire piece. You will not subsequently be satisfied with it. Be critical in your mind of the piece in the shop. Do not purchase until you are completely satisfied with every replaceable part—or its authenticity. The dealer will not be very pleased if you complain later. You must particularly beware of confidence in high-price labels. I have seen a blatant reproduction piece in one of our

20

glamour areas, priced at £95, which could be bought off the shelf from the supplier for about £5. Be your own judge.

One may start with some casual possession, added to equally casually. Soon somehow you will want information; interest will breed, and as knowledge grows, so will the quantity. For ideal display, pieces need ample space individually (Plate 1)—but successful collectors acquire less and less space. Accommodation is

Fig. 2
Overcrowding detracts from the overall appearance, and the importance of each piece: yet the pieces are too rare not to be on show.

limited, and so is the patience of the décor-minded collector's wife. Reference to photographs of any collection exemplifies the point, and one can only say that some collections are more so than others. Do try, however, to be selective and show as few as you need. Try to weed out the unnecessary. A shelf is an excellent display stand, or ideally a dresser for plates, measures and small pieces. Spoons (which are somewhat rare) can best be displayed on a spoon rack, and while originals are very scarce and expensive, reproduction racks are quite acceptable: or make your own

21

display shelf, preferably to show the knops standing up (unlike the antique racks).

Although many talk of a piece as 'magnificent' because it is huge, does attraction lie in vastness? Are not the smaller pieces more attractive? I would rather have a gill-sized hammerhead than a gallon. There are only two or three conventions in display, which are in any case normal artistic layout. Where possible have your larger pieces lower down; handles should be to the right for a right-handed person (although it is acceptable to balance a range of, say, buds on the right by double volutes on the left, with the handles of the latter to the left). Pieces with handles are best displayed with the handle about 45° rearwards, to show off the thumbpiece and a little handle. It seems quite hopeless to try to keep pieces of a close period together—it is far better to display aesthetically. On the other hand, there seems more than a psychological inhibition against mixing church pieces with other hollow-ware.

For safety, ease of identification, exchange of information by post, and other reasons, it is very advisable to photograph all your pieces. Also it is very useful and interesting to build up your own loose-leaf catalogue of your collection. Allot each piece a number, and record in its page all details of acquisition, whence, price, technical and subsequent notes, and attach its photograph(s).

Good luck to your collecting. I envy with vivid memory the flirtation and romance of acquisition and experience walking hand in hand.

3. Cleaning and Repairing

A great many people reading this book will own a few pieces of pewter, and to judge by the questions asked by individuals, the question of how to clean pewter, or whether to clean it at all, ranks as their biggest problem, greater even than knowledge on the various kinds of marks which are to be found on pewter. Therefore, to help those possessors in their logical sense of priorities, we are taking the problems of cleaning early in this book.

Cleaning

The very title of this chapter will cause some collectors—a minority —to bristle. There has been an oversimplified rumour around for a great many years that pewter should not be cleaned. Let us look into the pros and cons. Pewter was made to look as similar to silver as possible. It was finished extremely bright and highly polished. It is possible that hygiene was tacitly taken as assured with cleanliness, but more likely that scouring was the order to make it as shining and attractive as possible. Is it now fair to try to make a tricentenarian ape youth ? I think not—not to strip it bare of signs of age, then to buff it up to a very high polish. I know of no collectors who prefer this extreme. In this country it is thought to be

Fig. 3
The crusty surface of pewter when the oxide is completely dissolved. To rub down smooth takes great patience.

garish and unseemly. Commercially one is told that demand has it so, and it is indeed the easiest way to treat pewter. The pieces are immersed in strong solvent chemicals—either hydrochloric acid (spirits of salts) or caustic soda. The heavy unattractive dark scale is violently expelled off the face of the metal, leaving a very rough surface. This is reduced and smoothed on abrasive wheels, followed by buffing and polishing. Apart from bumps and dents all evidence and appearance of age have gone. It would seem to be reasonable to want each detachable part (lid, knop, thumbpiece, handle, body) to bear some trace of authenticity. At the same time, leaving traces of the black scale in all mouldings, interstices, spaces between rings and reedings, gives a very pleasing contrast of black and reflection (Plate 2). By leaving some scale on each part one has almost incontrovertible evidence of age. I say 'almost incontrovertible', because it is possible to fake scale with the aid of chemicals or use some other means to accelerate the formation of oxide. Furthermore, one must consider to what extent true oxide is a 'guarantee'. It forms quickly on some mixtures, slowly on others, quickly under some conditions, slowly under others.

To return—we are dealing with cleaning pewter. If it comes to you in fairly bright condition it is easy to keep it so. Any good metal polish (I find 'Glow' excellent) used twice a year is adequate, even in the smoke-filled centrally heated double-glazed confinement of the modern home. You may wish to use it more often. To sum up—it is not difficult to have a bright, lively attractive room display, easily kept in condition.

It is more difficult to put it into the condition most collectors like, in which your pieces will show honest signs of age, but will also be bright. We will come to that later in this chapter. First we must look into the very understandable objections to cleaning.

A minority of collectors prefer to keep every vestige of evidence of the passage of years carefully preserved on pieces. If they have acquired a heavy oxide scale, it is felt that you compromise their antiquity if you interfere. To the purist, a face-lift is not *comme il faut*. Some want to keep the heavy scale as proof of age. Others prefer the glamour and mystery of only half-seen beauties. And others, of whom I am one, leave pieces of great 'archaeology' or pre-pewter history uncleaned, so that they remain untouched for future research.

The oxides on tin and silver can be converted back to pure metal by modern methods, but it is possible that in the future research will enable the same with lead, and even tin and lead alloy, with the help of new techniques. Where there are pocks and breaks in the surface one cannot expect miracles.

Furthermore—and it is a very difficult decision in either case— (a) where there is delicate decoration (e.g. wriggling) or a weakly struck mark, will you lose it altogether if the scale is removed ? or (b) is the surface so unstable with pocks and eruption that removal of the scale will allow these conditions to become worse ? Until you have considerable experience in either of these cases, don't be impatient. You will be well advised to leave descaling for a year or two.

Finally, there is a very practical and human reason why you should not remove this really rather deadly, unsightly mask—it is a tedious, beastly, messy and energetic job to do, if the results I seek are to be achieved. You can always bring a pleasing vivacity to heavily scaled pieces with a couple of applications of furniture polish.

Just what is this dark oxide, and what are the risks of losing marks underneath it ? If pewter is left untended and exposed to alternating conditions of high humidity and dry air, plus other oxidising conditions such as natural deposits of dust (see how surfaces facing upwards such as lids are worst affected), then first a 'hume' forms, then a grey film and tarnish appear. This is caused by oxygen attacking the baser of the two metals in the alloy, and causing a skin of oxide. Cotterell in *O.P.* was of the opinion that low temperatures accelerate the formation of scale, but I think that this is a misconception and a reversion to the old misno-mer 'Tin Pest' to which we will refer later. It is very important to appreciate that (with the exception of pocks) the surface of the scale appears to be formed very accurately over the original surface, incorporating the finest detail of weakly struck marks, or the faint lines of decoration. But underneath, the surface of the metal is broken up more and more irregularly as the scale thickens.

Thus, in a nutshell, if you remove a thick scale completely, a very rough, crystalline surface of the alloy awaits you, comparable with the appearance of snapped cast iron. This very irregular surface has bitten and detrited perhaps every trace of your weak touch mark—so note the detail *before* you attack your piece aggressively. This oxide is more apparent in lower-grade alloys—and it is far heavier on some nineteenth-century pub tankards than on some seventeenth-century and earlier ware. So you can see that there are problems which you must appreciate before you begin to tackle them.

The fine copper content pewter of the early Middle Ages is very free from dark oxide (but does sometimes bear oxide like gilding). To take an example from further back, a beautiful Romano-British pewter plate which I cleaned for Norwich Castle Museum needed only a minimum of treatment with very fine-grade emery. Its mint condition was probably due to the fact that the river bank in which it had lain buried for hundreds of years was chemically neutral.

As well as the all-over formation of scale, I have mentioned pocks and eruption. For some reason, probably impurities in the alloy, the oxide sometimes forms in pocks more deeply, at a greater speed, and as a result is softer. These pocks are very unsightly and it is easy to see that chemical means of removing the flat scale will attack the softer oxide more strongly than the rest, leaving deep pits. I recently acquired a flat-lid flagon whose

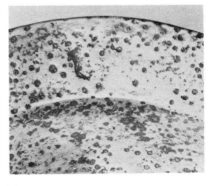

Fig. 4
Pocks are caused by impurities in the alloy. The oxidisation bites deeper in some places and, having formed more quickly, is softer, and therefore dissolves more completely than the average surface. (See Fig. 6.)

26

previous owner, with the best of intentions but the minimum of experience, had dissolved off all the scale. It looks as if it had been hit with pellets from a sawn-off shotgun. If only he had left it for more experienced hands.

An even more insidious condition is eruption. Sometimes you can see bubbles of pewter appearing, although there was perhaps no sign of them before you started to attack the scale. A few months after descaling, the surface may slowly quake up little domes of pewter about 1 or 2 mm. across, very reminiscent of horrible bubble gum. In this case it would seem that removal of the scale

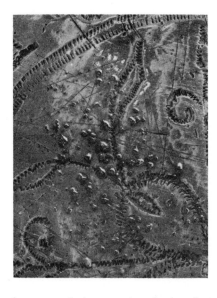

Fig. 5
Sometimes removing the scale allows further oxidisation under the surface, causing it to erupt. There is no known method of neutralising this condition.

has unsealed an under-the-surface cause of oxidisation. We do not yet know a method of neutralising it. The remedy, if you are a patient, is to drill out each 'volcano' and replace it with matching metal.

These conditions of bad scaling, pocks and eruption have been confused with the term 'Tin Pest'. I am sorry and surprised to see Michaelis perpetuating this term, for surely Cotterell in his O.P. had killed it, as far as pewter is concerned; and it has long been accepted that *all* authorities use the term for tin ONLY. 'Tin Pest' is a very rare condition, where occasionally pure tin which has been subjected to lower than normal temperature 'perishes'—loses its crystalline structure and cohesion. If it is in this condition and is disturbed it is liable to fall into a heap of dust. Furthermore, for some extraordinary reason, it is contagious—contact with a sound piece at subnormal temperature (say less than 55° F.) can pass on the infection. I understand that if it is undisturbed or undamaged such affected tin can be restored to full structural strength by heating. I may add that I have kept some pure tin in my deep freeze for a long time, hoping to witness the phenomenon —but my tin is very conservative and unco-operative. The leading authorities on tin confirm that this condition applies *only* to tin and *NEVER* to pewter. Alloying stabilises tin against it occurring. Let us hope that never again do we hear 'Tin Pest' applied to pewter. The full effect of oxide on pewter has never been fully examined under laboratory conditions. As regards Britannia metal, the tarnishing and oxide are very much harder, and more difficult to remove. The lead constituent of pewter is replaced by antimony— and the oxide formed by antimony is very 'tight'. Also, since Britannia metal is hard, one is reluctant to do much work with abrasives, which could cause long-lasting scratches.

By this time our possessor of a few pieces, blinded with much non-applicable science, is no doubt getting worried and he probably decides to use only furniture polish. But he may be anxious about how best to take further steps. The short answer is, the less the better. Use the least aggressive methods possible. I do not mean that you should rely on Cotterell's extraordinary support for 'boiling in hay'. By all means try it. I think the only outcome is that you will need to boil it for so long to get any result that all the water will boil away, leaving in the bottom of the saucepan a little pool of bright molten pewter which was once your flat-lid tankard. Perhaps my hay, or my imagination, is not sufficiently strong.

If your pewter is already clean, then good metal polish, such as 'Glow', is adequate. I think Cotterell's suggestion of using oil is bad, despite its sealing effect. Dust will adhere to the sticky oil, and dust is notoriously corrosive.

If there is only a very little scale, and if you are sure you can cope with it, then use the very finest grade of wet-or-dry emery paper (600 or 800) used wet, with the initial cut of the abrasive taken down by rubbing it on, say, a screwdriver handle. Rub the piece in different directions. It is as well to stop before you have gone too far—you can always return to do more. I advise against using wire wool in any form, because some of the wisps of wire will lie along the direction of rubbing, and will cut deeply. It is a good idea, when dry, to rub vigorously with leather 'split', to smooth

28

the surface. I use 'split' leather mounted on a flat stick 18″ long. This is to me the most important tool in cleaning, but it can be used only on outside curves. This rubbing down is hard work, for you need to bear on it hard, to bring up a fine texture. Britannia metal, being harder, is less amenable to one's efforts with emery, and you will have to work carefully to avoid scratching it. After rubbing down, use metal polish, of course. The advantage of a metal polish with 'body' is its tendency to fill scratches.

May I suggest that you read all the chapter before rushing into any action other than the use of metal polish. You may be tempted to use some kind of mechanical means—a hand power tool or a static polisher—to get rid of film or oxide. My advice is, Don't. With the former tool it is difficult to mount sufficiently fine emery paper; you have to work dry, and with one hand; and you are very liable to lose fine control and so bear into the piece—perhaps with the central mounting: or it may 'chuck'. The static polisher does allow use of two hands, but it is very difficult to get to all the parts you want with the proper pressure. Speedy pressure, too, can 'burn' the metal. It is far far better to do it the long way, purely by hand.

If you are faced with a piece with more scale than you judge you can remove with 600 emery paper, you will need to use chemicals —either hydrochloric acid (spirits of salts) diluted or caustic soda. I have used both. They are equally effective, but hydrochloric acid is more difficult to rinse off, and therefore harmful salts can be left in cracks and crevices in the metal, storing up trouble for the future. Some prefer one, some the other—my own preference is strongly for caustic soda and so in fact is that of most of my friends. Both are highly corrosive—murderous on clothes and carpets, and satanic on the flesh. First clean off any wax or polish, then, wearing rubber gloves, with water in an egg cup, drop in the caustic soda, stir to dissolve, soak a rag, and sponge over the surface to be treated two or three times, then stand it on news-paper.

Earlier we discussed the advisability of leaving some scale on every detachable part of a piece, and of leaving black scale in the moulding, etc., to throw up a light and shade. If you strip a piece bare it is not only garish, but you have removed vital evidence of its age (Plate 3). So it is well, and attractive in appearance, to leave some scale in all mouldings, fillets, reedings, angles and under the lee of other parts, e.g. on the drum next to the handle, etc. You can smear Vaseline on any part you want to protect.

29

Better still, for more delicate work, e.g. in the lettering of engraving, or wriggling, I use black shoe polish. You may want to leave a mark unattacked—just smear the protective coat over it. Little and often is the key word for application, each time using some 400 wet-or-dry emery, with water, to help ease off the scale. It is not as easy as it may sound. The scale is thicker in some parts. Dents are more difficult to cope with. Some patches come off easily. I can only leave it to you to be patient and persistent, and gain experience. It is a very long job if you want to do it well—but most satisfying.

Now for the emery; you can use the 400, rubbed down, for the start, because, as mentioned previously, the metal surface will be very rough. (Actually, since the oxide has formed on the lead content, the skin metal left will be tin.) The 400 will leave very obvious scratches, which will be partly removed by the 600, and improved still further with the leather stick. Finally the polish, when rubbed off, will show you what you have achieved, and if it is satisfactory—most probably you will need to return to some parts at least.

We have talked of the evil of pocks. It is particularly dangerous to attack the scale too viciously in case there are deep soft spots. It is far, far, better to keep some patches of scale than to have deeply corroded bare patches. So, wherever feasible, always use emery paper rather than complete removal. If a piece is one, two, three hundred years old it can wait a week or two while it is treated with loving kindness rather than brash hastiness.

A week or two? Obviously, people who make a business out of buying and selling pewter cannot afford to waste so much time on one piece. It would make the cost of a nineteenth-century pot too high. So the trade strip and then buff. You should be able to buy pieces uncleaned, which you can treat properly yourself.

It should be mentioned that some pieces can be treated not only with caustic soda, but by sprinkling granulated zinc on the surface of the piece which is still damp with caustic soda. This causes a chemical electrolytic action, but it is rather fierce, so it should be used with great care on *stubborn* patches. And that there will be some is certain. Cleaning your pewter, and handling it, will teach you a great deal about the metal and make you wonder how the piece was made. You will soon become aware of the seams running round pewter, and down Britannia metal, for instance; you will start to wonder how handles were attached and how pewter was assembled. Briefly, all pewter has always been cast from moulds.

30

Bath Museum has a number of examples of stone moulds which were used for the earliest pewter—that of the Romano-British period, made nearby in about the fourth century. Coarse stone was used for dishes, and lias for small delicate pieces. Since the Middle Ages the moulds have been made from bronze. There are many moulds in existence dating from the seventeenth century for various types of ware—plates, spoons, jugs. Those for jugs are more complex, having not only the protuberances such as handles, lids, thumbpieces, etc., but also the complications of separate moulds for the upper and lower parts of bowed-out curves—i.e. stouter in the waist than at the neck or base. (How else do you remove the inner part of the mould?) The alloy was melted and the mould heated (to allow the molten metal to flow freely to every part). When removed, some ware was hammered to condense and harden the metal, and all circular parts were cleaned up on a wheel. The turning marks are very clear on Romano-British dishes and on the bases of seventeenth- and eighteenth-century hollow-ware (measures, tankards, flagons, etc.). The various parts were then assembled. Moulds, as well as being extremely heavy (those for a pint jug weigh 60 or 70 lb.), were very costly, and some were kept by the Company and loaned or hired to smaller makers. As it is also possible to cast pewter in sand and plaster, one wonders to what extent, if any, these were used. Presumably fakes are made in plaster models taken from originals. After the parts of the body were assembled, it was trimmed up with a wheel or lathe, then rings, if any, would be incised.

Britannia metal is 'built' in the same way as silver—it is spun, i.e. sheet metal is laid over a mould on a lathe and is then bedded down to it by pressure as it rotates, the seams being soldered later. Whereas we categorically stated that all pewter was cast, there is a slight complication in saying that all Britannia metal was spun. First, a piece with considerable curves might be spun in two sections, and the equator seamed; second, Britannia metalware often has handles of cast pewter, and if it has legs, they are always of cast pewter. Some other small parts may also be of pewter. These parts are burned on. No parts of either pewter or Britannia metal were riveted, but in late wares the knop on lids sometimes had an embedded bolt.

On acquiring pieces of pewter, there may be many dents, and the rim of a plate may resemble the brim of a wet felt hat. Two forms of very elementary restoration present themselves—to tap out the dents, and to bend the waves straight. Tapping must not be done with a hard metal hammer. It will damage the pewter. Even a brass pestle with a rounded end makes very sharp dents. Use a hard rubber hammer. Or a round-headed hammer cast by yourself of pewter or lead is ideal. Always tap from the inside against a surface, not forgetting to turn the body of a curved piece so that the small area tapped has a firm anvil. Take great care not to distort by over-tapping. With a metal as soft as pewter, it is only commonsense to hammer lightly and very frequently—in bottom gear; not heavily and top-geared. Tapping may fracture the scale and cause it to flake off (Plate 4)—so you will have to decide whether it is worth removing the dent at the risk of damaging the scale or not. Many tankards, plates, etc., will be misshapen, but can, with patience and care, be worked back into shape by hand— forefingers, thumbs and the ball of the thumb. It is far safer to do a little at a time, and allow the metal to settle and readjust. As with tapping, only a small alteration to the shape may cause the scale to flake off complete. Sometimes, on battered wells of pewter and dishes, I take my shoes off, and work the flat of the well into shape with my heel and ball of the foot on a smooth floor. But first one needs to be certain that the well should be flat—not all types were! I find it a great help to immerse a piece for a few minutes in a saucepan of hot water. You will find that raising the temperature to boiling point makes the metal more malleable. Then, wearing gardening gloves, one can work the slightly softened metal more easily while it is still hot. This is particularly useful, even important, in working solid parts such as thumbpieces which have been crushed. It is fatal to try to re-shape them too quickly.

Repairing

Repairing is a most satisfactory craft to master—but most frustrating to attempt and not master. It is a skill, it appears, for which one either has or has not the gift. If you are, like me, one of the many who seem to be considerably bumblefisted at delicate jobs, probably with an inherent optimism and impatience that by-passes adequate preparation, then I can only recommend innumerable practice joins and dummy runs on scrap pieces. They should show 100 per cent success before you risk tackling a favoured piece. Don't forget, too, that it is not easy to mock up the repairs to simulate the real difficulties you will face when you come to the real thing—the rim separated from the lower part of a plate, a leak in the mouldings at the base of a tankard, or the replace-

Plate 3 Dissolved and stripped, the appearance is garish and lacking in personality.

Plate 4   If a heavily scaled piece is bent, the brittle oxide will flake off, leaving a most unsightly contrast.

ment of a handle. Furthermore, each repairer develops his own quirks of technique.

Perhaps one should mention that some minor repairs can be tackled while you are still not an expert, but nevertheless in a practical manner, by using epoxy resin and polyester adhesives. No heat is used in this treatment. One adhesive which is metal-filled and quick-curing—as little as 5–10 minutes—gives adequate results on pewter of no great value. It dries to a powdery grey colour which does not match (so repairs should be out of sight) and if not perfectly mixed is suspect when immersed in hot water. I have used it with complete success to replace handles and fill minor holes. I have even backed up small holes in wafer-thin sections of pewter with very small patches of a fine linen handkerchief soaked in the resin. It is, however, difficult to judge the amounts required in a mixture of two ingredients of differing proportions, e.g. 10 : 1. These adhesives result in a repair that you want to stay hidden—not to advertise your technique!

Prestige does follow skilful metal running, using either a bit or blowpipe. Your bit should be large enough to hold the heat from dispersing too quickly, the temperature as low as you can keep the solder running. For solder, it is best to use pewter or Britannia metal, as this will give a closer colour match. Meticulous cleanliness of the bit and the surface to be joined is absolutely essential. Conversely, you can limit the area of adhesion by not cleaning the unwanted area. Shape the damaged part together, and then prepare a new gully of V section with a file. It is presumably unnecessary to elaborate on the use of flux, and tinning the bit. Dipping the bit into your solder, take up a little, keep the head of the bit down, and drop some into the crack. Never mind if you add too much—you *must* add too much to allow for taking down exactly to the surface, by power tool, dental mechanic's accessories, or by file and scraper and emery cloth. Run it in, building up until the gully is filled. You may become so expert at handling the bit and judging temperature that you will be able to do the whole repair in one stroke. Similarly, try to learn where the heat of a blow flame really is, and then experiment with several types of crack or hole.

Patches to replace holes must be carefully curved, if necessary, before applying the solder, and should be visibly smaller. Attach one point first, positioning perfectly, then fill in each corner, then the whole. It is essential to maintain the contours of the piece to be repaired. Your tools will allow this, provided you use plenty of patience and don't remove too much. You can always take more off later. Naturally, one needs plenty of substance of metal (not including oxide!) to absorb and disperse the applied solder. If you add a mass of molten metal to a wafer-thin part, the latter will disintegrate.

For badly pocked or erupting pieces it may be necessary to drill out the sources of trouble (usually if there is one there is a multitude) and to solder in discs of pewter. If they are individual holes, obviously it is necessary to drill out and fill. It is all very much like dentistry, but with much more to fear! And not nearly so fast. Cleaning out the blips of solder inside the drum, or under a dish, can be partly achieved by touching with a hot iron. If the sur-

Fig. 6
A fine Tudor plate which should not have had the oxide removed, since it is very pocked. The metal remaining after the unwise stripping is very thin. Note the now exceedingly rare example of the Rose and Crown stamp of quality. No touch mark visible. Diameter $11\frac{1}{4}''$. (See Fig. 4.)

rounding metal is not clean, the re-molten solder will not adhere to the scale or dirty surface. Finishing off is done by using decreasing coarseness of emery paper or cloth. The part repaired will now be a very bright patch. This can be toned down by applications of diluted nitric acid (50 per cent) which stains the metal grey.

It is sometimes necessary to make a new part to replace, say, a missing thumbpiece. First find a model to reproduce. (Note carefully the correct angle at which to attach such pieces.) Take a plaster-of-Paris cast, adding a 'handle' to the part so that you can grip it in a vice for subsequent work. Cut your cast open, make a hole for filling, and then run in your molten pewter. Proceed as in repairing, setting the attachment in position by one easy adhesion, then fill in. Finally, clean up, and stain to darken. For lids—you may be able to turn the whole or parts of a casting on a lathe. It is very difficult to space the casting sufficiently accurately, and so for

34

the substance lids it is much simpler to turn the surplus away, than to try to achieve such accuracy in a mould.

At all stages ingenuity, patience and deftness are essential. Practice on scrap or worthless pieces. Do not be too ashamed of not completely trimming away all the surplus fringe—experience of old pieces shows that very seldom was the job mechanically perfect. Herein lies the charm of hand work—the irregular minute imperfections.

4. Marks

We have seen how the control of the pewterers was completely in the hands of the Pewterers Company, occasionally reinforced by Parliament, and that their control and checking were jealously guarded—certainly up to 1550. From then on there were increasing creaks and groans, developing into dissension. One would have thought that what was established as a craft by the mid-fourteenth century, and received its charter in the fifteenth century, would by that time have instituted a method of checking back—of identifying the makers by some means of marks.

Not one single piece of Romano-British pewter (except possibly one, from Cirencester) bears any form of maker's mark. But it is more surprising to find that none of the earliest-known medieval pieces is marked—the sepulchral chalices buried with monks in the twelfth, thirteenth and fourteenth centuries. Nor are the few reasonably dated medieval flagons or plates. This throws into new light the Act of Parliament of 1503, whereby the marking of pewter was made compulsory—perhaps for the first time. Scotland followed in 1567. Presumably the Company's searchers had had to catch the quality evaders redhanded. Perhaps they had long been enjoying trouble-free trips out into the country. They should have been seizing faulty wares, imposing fines, and, in the later fifteenth century, marking faulty pieces with a broad arrow. No such pieces are known, due chiefly, not to normal wastage, but to the fact that these pieces would have been melted down to be recast. With no means of identifying wares, the searchers must have had an uneasy sinecure.

The Act of 1503 changed all that. Henceforth every piece had to be marked. (There had just been a reference to the Company buying marking irons for the hollow-ware men.) The evidence is most easily seen on spoons, the most numerous kind of ware still existing. In the fourteenth and fifteenth centuries they were seldom, if ever, marked, whereas nearly all sixteenth-century spoons bear delightful, excellently made touches. But it is not until 1550 that we read of a tablet on which every (?) man's touch was struck. It seems highly unlikely that all or many provincial pewterers came to London to 'touch' the tablet (or touch-plate). We know very many later provincial pewterers who for one reason or another did not strike their touch mark. But despite the implied guarantee of quality given by signing pieces, the troubles with debased alloy, trickery, poaching and so on increased. We cannot hope to see this touch-plate, for it was lost in the Fire of London in 1666. A new touch-plate was immediately put into use, and every pewterer was required to restrike his touch. Cotterell in his *O.P.* unravelled the seemingly anomalous sequence, pointing out that, after the Master and Wardens, they struck in seniority, to be followed by the new brethren as they were given leave, and before they could set up in business. However, there are many, many marks which are missing from the touch-plates, but which are recorded by Cotterell, and there are many which did not come to Cotterell's notice. Many of these are recorded in private hands, and it is to be hoped that the compiler will soon publish them.

Seventeenth-century pieces of beautiful technique, metal and design are often found unmarked. Why the makers were so shy of putting their touch on their wares is not known. Were they in trouble? Were they not qualified, making pewter pieces as a side-line? Were they too poor to buy an iron? Unlikely types of ware were unmarked—for instance it is most unusual to find one of the magnificent and numerous, always excellently made, 1605 type of church flagon with a touch. There are several records between 1520 and 1670 of fines for not marking pewter-ware (including one for $2\frac{1}{4}$ cwt. of tavern pots (!) in 1595).

So, as with present-day shopping, judge the piece, not the label. It is sometimes possible to deduce the source of an unmarked piece, if it be of a rare type matched by pieces bearing the touch of an identified maker. The only other two centres known to have had touch-plates are York, whose plate has long since disappeared, and Edinburgh, whose two plates still exist. The Edinburgh marks range from c. 1600 to c. 1760. All the marks are very similar, consisting of the three-towered castle with makers' initials, and sometimes the date of striking.

Welch delved most thoroughly into not only the Company history, but the history of marking pewter, which is admirably laid out in both 'Cott. O.P.' and Michaelis in his *Antique Pewter of the British Isles (A.P.B.I.)*. They trace the growing unrest within the Company, and show how, beset by surges from below, intensified by competition from other media and vacillation by the Company, all control was lost by the middle of the eighteenth century.

We have seen that irons were provided in the last quarter of the fifteenth century, and that marking of pieces was made compulsory in 1503. Constant friction developed throughout the century, including makers encroaching on one of the few marks placed on pieces by the Hall—the Rose and Crown (in that century), which, although records are confusing, appears to have been a mark struck on goods for export—an emblem for 'Made in England'. Those who offended had their goods seized, were fined or were required to strike a new touch. Presumably this was intended to let them turn over a new leaf, and not be dogged by their past faulty wares. Competition arose from stone pots, and the Company tried all kinds of devices to suppress this new threat, even to petitioning that all measures should be made of material which would take the impression of a seal. Perhaps that was the origin of the pewter lids on stoneware pots known to have been made here as well as in the Germanic countries. (Their descendants are made to this day in Austria and Germany.) In the seventeenth century the old complaints reappeared, and after the Fire there is much evidence of emancipation of thought and deed. The spoon-makers, a group on their own, seem to have been up to some trickery, for in 1666 they were all ordered to have new touches. There were also more complaints of anonymity.

We will deal later with the form of the marks, but will now touch on their content. Early marks had appeared to be simple and direct—just two initials. But from the early years of the seventeenth century symbols begin to appear. They indicate the name or address of the maker. You do not have to guess long at the name of the maker whose touch is T.B. and a bell. Others are more punning and less obvious; Lucas shows a shaft of light (Latin *lux* =

Fig. 7
An early touch mark on the front of the rim of a small early sixteenth-century plate—'D.I.' Diameter $7\frac{1}{2}$".

light). This was the fashion of the period, and continued in the next century. The copper token 'coinage' of 1649–72 abounds with this same pictorial representation of name and address, which was read so easily by the illiterate. Our company symbols today are similar, but much more abstract.

Perhaps the pewterers took a lead in self-advertisement from the token issuers. In the past this was just 'not done'. But trade was bad, and, to help to maintain business, makers now put their names in full in their touches, added 'London' and their addresses, and finally a sales message such as 'Superfine Hard Mettle', etc. The provincial pewterers, to the annoyance of their citizen brothers, also put 'London' on their wares, and, as it turned out, there was nothing to stop them. In the meantime the London makers again encroached on the use of the Rose and Crown, and adopted it. The Company has lost control of them. In the early eighteenth century there were again complaints of the state of trade; insolvencies were rife, and with fifty-odd years of vacillation and ineffectiveness, events proved stronger than the Company.

The time of all this trouble—primarily 1666–c. 1705—when distress and unrest ruled, is the very period we regard as the 'golden age' of pewter. It almost seems as though all the most desirable ware was made in this short time, and what a quantity is to be seen in collections! Certainly it is the high period for the combination of design, decoration and workmanship.

There are several other types of mark which each individual pewterer used, broadly classed as secondary marks, including 'hall marks', cartouches, Rose and Crown and X crowned. These are dealt with later in this chapter. There are also capacity-checking marks. But the touch is the basic vital factor of interest, so let us see who was entitled to strike his touch.

39

There were five stages of progress in seniority of membership, or brotherhood, in the Pewterers Company. At first a boy was apprenticed for about seven years, learning the trade thoroughly, with long hours for little pay. He lived in with his master, and was probably exploited to the full. At twenty-four he was eligible for the next step—admission to the Company as a Yeoman—the Freedom of the Company. If he satisfied his seniors sitting in judgment as to his workmanship (for he had to submit an example of his work) he could be given leave to strike his touch. He would almost certainly do so only if he wanted to set up on his own. But he might not have sufficient capital, so he would work for another as a journeyman on a higher status than an apprentice. (A journey-man was an artisan, not a traveller.) Let us say that he wanted to set up on his own. He designed and procured his touch iron, and in the presence of the Master and Court struck his touch on the touch-plate.

If in a few years he had been successful and had built up some capital—the required sum was as high as £500 in the distressed golden age at the end of the seventeenth century—he might be called to the Livery of the Company. Probably the strength of the Livery was kept to a fixed number, but the strength of the Yeo-manry was flexible, influenced by economic factors. It was prob-ably only as a Yeoman that he was called on to take up arms to help quell disturbances, and he would be relieved of this duty on becoming a Liveryman. As a brother on the Livery he would sit at Common Hall on matters on which the Court wished to consult the whole Company.

As his seniority increased, so in time he would be called as an Assistant on the Court (the management committee). If he proved business-like he would be elected Renter Warden, and then Upper Warden, a year at each, probably soon to be followed by the honour of being Master, which, despite the democracy of the Common Hall, was a position of considerable power. Understand-ing the organisation, particularly of the point at which our pew-terer was allowed to strike his touch, enables one to appreciate one of the reasons why so many Yeomen are shown in 'Cott. O.P.' for whom marks have never been found. They may never have opened up on their own, and therefore had no need of an iron. As already suggested, perhaps they were provincial and did not travel up to strike their touch; perhaps they died before making many pieces, none of which has survived.

Each maker's mark is known to collectors by two different

numbers, if they are struck on the London or Edinburgh touch-plates. Otherwise they are known by one number: e.g. (a) 'L.T.P. 1021' being the 1021st touch struck on the London touch-plates (of which there are five, with a total of nearly 1100 touches); (b) if recorded by Cotterell, as 'Cott. *O.P.* 2440'. Naturally he records all those on the touch-plates. Where known to him, Cotterell gives details of makers. Let us look at his entry for John Hudson, whose biography is exceptionally full. Explanations are given in brackets.

> '2440. Hudson, John, London: 22 Mar. 1770, Y. [admitted to the Freedom as Yeoman]; 23 Aug. 1804, L. [called, or 'clothed' with the Livery]; 1808, f.S. [fined for not serving as Steward]; 1818, R.W. [Renter Warden]; 1822, U.W. [Upper Warden]; 1824, M. [Master]. Died 17 Feb. 1829. Touch, L.T.P. 1021, which he obtained leave to strike on 20 June 1771. In 1793 and 1801 he was at 41 Fetter lane.'

Then is shown his touch mark, and two subsidiary marks. Other entries may have far less detail, and perhaps no mark is known. Cotterell decided wisely to compile his list of known pewterers alphabetically, for easier cataloguing and reference. It would be difficult to arrange them by dates, as so many are conjectural, and reference would always be tedious. Confining a section to London touch-plates would duplicate searches. Alphabetic sequence keeps together members of the same family who sometimes used very similar marks. Uncertain or blind allocation necessitates a separate section, and unallocated marks with no initials a further section. Not for the first time, may I say that Cotterell should be read! I am grateful to Messrs B. T. Batsford Ltd for permission by arrangement to reproduce a small selection of some of the marks which the collector of today is quite likely to encounter, these having been struck from c. 1750 onwards. The numbers quoted are 'Cott. *O.P.*' numbers, and the entries in this book are simplified to give what is known of the span of that mark, to help in dating pieces. This book does not set out to be a complete reference, but to be a really useful companion in the car as well as at home.

38. Alderson, Sir George, London: July 1817, Y.; 21 Aug. 1817, L.; 1817, f.S.; 1817, f.R.W.; 1821, U.W.; 1823, M. Sheriff of London 1817. Died 1826. Touch, 1084 L.T.P.

40. Alderson, Thomas, London: c. 1790–1825. He made the pewter-ware for the Coronation Banquet of George IV.

118. Ash & Hutton, Bristol: c. 1760. Partners, Gregory Ash and William Hutton.

153. Austen, Joseph & Son, Cork: Were at 54 North Main Street, 1828–33.

250. Barber, Nathaniel, London: 16 Sept. 1777, Y.; 20 June 1782, L.; 1787, S. The *London Gazette* records his Bankruptcy on 22 Mar. 1788. Touch, 1037 L.T.P. Was in business in Snowhill. (Compare the Touch of John Home, whom he probably succeeded.)

407. Bentley, C., London: c. 1840. Of Woodstock Street.

430. Birch & Villers (see also Yates & Birch), Birmingham: c.1775–1820.

532. Bowler, Samuel Salter, London: 18 Mar. 1779, Y. Touch 1038 L.T.P.

574. Bright, Allen, Bristol and Colwell, Herefordshire:
2 Nov. 1742, F. S. of Henry. Apprenticed to William
Watkins, Bristol, and Mary his wife, 1 Nov. 1735.
Premium, £45. Died in 1763.

708. Burgum & Catcott, Bristol and Littledean, Glocs.
Partners, Henry Burgum the elder, and George Symes
Catcott. Partnership was dissolved on 16 Feb. 1779.
(*London Gazette.*)

740. Bush & Perkins, Bristol and Bitton, Glocs.: Partners, Robert Bush, above, and Richard Perkins, q.v. c.1775.

873. Chamberlain, Thomas, London: In the lists of Openings and Touches, leave was given to Thomas Chamberlain to strike on 20 Mar. 1734. 3 Aug. 1732, Y.; 24 June 1739, L.; 1751, f.S.; 1754, R.W.; 1764, U.W.; 1765, M. His Trade-card, q.v., gives his address as King Street, corner of Greek Street, St. Ann's, Soho. He is mentioned until 1806. He was later a partner in Chamberlain & Hopkins.

962. Cleeve, Bou(r)chier, London: 16 Dec. 1736, Y. and L.; 1744, f.S. and f.R.W.; 1755, f.U.W.; 1757, f.M. Was given leave to strike his Touch on 22 June 1738. Was in partnership with Richard Cleeve, vide Welch, ii. 194. Lived at Foots-Cray Place, Kent.

1004. Cocks, Samuel (see also Cox), London: 11 Mar. 1819, Y. and L. Touch, 1080 L.T.P.

1063. Compton, Thomas, London: 5 Aug. 1802, Y. ; 24 Sept. 1807, L. 1817, died. Succeeded to the business of his father-in-law John Townsend who died in 1801, to whom he was apprenticed in 1763 and whom he joined in partnership in 1780. He was succeeded by his second son, Townsend Compton, in 1817. In 1793 his address was Osborn Place, Brick Lane.

1360. De St. Croix, John, London: 11 Dec. 1729, Y. Touch, 833 L.T.P., which he had leave to strike on 18 June 1730.

1466. Duncomb(e), Samuel, Birmingham: c.1740–1775 or
1780.

1477. Durie, ——, Inverurie, Aberdeenshire: His name
appears on Scottish snuff-mulls. Late eighteenth
century.

1547. Ellis, Samuel, London: 28 Sept. 1721, Y.; 1725, L.;
1730, S.; 1737, R.W.; 1747, U.W.; 1748, M. 1773,
died. Touch, 746 L.T.P., which he had leave to strike
on 10 Nov. 1721. He was succeeded by Thomas
Swanson.

1576. Englefield, William James, London: Was an appren-
tice in 1867 (with John Jarvis Mullens). 21 Mar. 1875,
Y. and L.; 1890. f.S.; 1907, R.W.; 1908, U.W.; 1909,
M. Touch, 1091 L.T.P. Englefield's business is of very
ancient foundation and traces its descent as follows:
Thomas Scattergood, 1700, Y. Succeeded by
Edward Meriefield, 1716, Y. Succeeded by
John Townsend, 1748, Y.–1766, then
Townsend & Reynolds, 1766–1777, then
Townsend & Giffin, 1777–1801, then
Townsend & Compton, 1801–1811, then
Thomas & Townsend Compton, 1811–1817, then
Townsend Compton, 1817–1834, then
Townsend & Henry Compton, 1834–c.1869, then
Elmslie & Simpson, c.1869–1885, then
Brown & Englefield, 1885–present day.
(The above notes were corroborated by Mr. W. J.
Englefield in the years immediately preceding his
death.—H.H.C.)

1740. Fothergill, M. & Sons, Bristol: Mentioned in Matthew's first Bristol Directory (1793–4) as of Redcliff Street. This partnership was dissolved on 17 Aug. 1805.

1909. Go(e)ing, Richard, Bristol: 7 Feb. 1715, F. Mentioned in the Poll Book for 1734 as of St. Stephens, Bristol. In an advertisement of 1742 he was described as of The Block Inn (his House) on the Quay, and as having been there for 25 years and using the Lamb and Flag mark with his name upon his wares.

1943. Graham & Wardrop, Glasgow: c.1776–1806. Partners, Robert Graham and James Wardrop. Both were admitted to freedom in 1776 as coppersmiths and white-ironsmiths, but much pewter-ware made by them has come down to our time.

2162. Harrison, John, York: Mentioned in 1712/13. 22 Feb. 1724/5, F. City Chamberlain, 1745/6, and was searcher in several years.

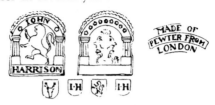

2177. Harton & Sons, London: c.1860–1890. Succeeded Watts & Harton, c.1860, and closed the business which was in High Holborn in 1890 when the manager and most of their connexion was taken over by Brown & Englefield.

2329. Hincham, A., of ?: c.1720–1750. (Cf. Touches of Watts & Collyer.)

2341. Hitchman, Robert, London: 16 June 1737, Y.; 11 Aug. 1737, L.; 1749, S.; 1752, R.W.; 1761, U.W. Touch, 877 L.T.P., which he had leave to strike on 11 Aug. 1737. In 1743 was at St. Margaret's Hill, Southwark.

2393. Home, John, London: Of Snowhill, 14 Dec. 1749, Y.; 21 Jan. 1754, L.; 1763, S.; 1771, R.W. Touch, 965 L.T.P. In 1754 he was allowed 'Mr. Warden Smith's Touch by consent,' but on 19 June 1755 he was 'ordered a new Touch as Smith still used the old one.' Struck c.1754. (Cf. Natheniel Barber, by whom he was succeeded.)

2750. King, Richard, Junr., London: 12 Dec. 1745, Y.; 18 Dec. 1746, L.; 1758, f.S.; 1763, f.R.W.; 1776, f.U.W.; 1777, f.M. Died in 1798. On 12 Dec. 1745 was granted leave to strike a Touch the same as his father's.

3147. Maw, John Hornby, London: 13 June 1822, Y. and L.; 1835, f.S.; 1847, f.R.W. Touch, 1087 L.T.P.

3153. Maxwell, Stephen, Glasgow: In 1781 is mentioned as Copper and White-ironsmith of Maxwell's Street, and in 1784 as a Pewterer.

3317. Moyes, J——, Edinburgh: in 1872 his shop was in West Bow. He was the last pewterer to practise the trade in Edinburgh.

3334. Munster, Iron Co., Cork: c.1858–c.1905. (Cf. Austen & Son.) c.1833 onwards.

3694. Pitt(s), & Dadley, London: Touch, 1043 L.T.P., struck c.1781. Partners probably R. Pitt, 1780, W.; and E. Dadley, 1799, W.

3732. Porteous, Robert & Thomas, London: Touch, 999 L.T.P., struck c.1762. Of Gracechurch Street. Successors to Richard King, hence the ostrich.

3817. Ramage, Adam, Edinburgh: In 1805 he was apprenticed to James Wright.

RAMAGE EDIN?

4379. Smith(e), Samuel, London: 17 Oct. 1728, Y. Son of Jno. Smith. Touch, 796 L.T.P., which he had leave to strike on 21 Mar. 1727/8.

4442. Spackman, Joseph & James, London: Touch, 1045 L.T.P., struck c.1782.

4795. Town(s)end, John, London: 16 June 1748, Y.; 21 Jan. 1754, L.; 1762, S.; 1769, R.W.; 1782, U.W.; 1784, M. Died in 1801. Touch, 928 L.T.P., which he was given leave to strike on 16 June 1748. He was apprenticed to Samuel Jefferys in Middle Row, Holborn. Commenced his business at 47 Prescott Street, Goodman's Fields.

4876. Villers & Wilkes, Birmingham: In 1805 Directory they are given as wholesale braziers, pewterers and dealers in metal at Moor Street.

4959. Wardrop, J. & H., Glasgow: c.1800–1840.

4968. Warne, John, London: Founded in 1796. Of Blackfriars Road. Now incorporated in Gaskell & Chambers.

4996. Watts & Harton, London: c.1810–1860.

5130. Whyt(e), Robert, Edinburgh: 1805, F. No. 40 Cowgate Head.

5338. Yates, James, Birmingham: c.1800–1840. Now incorporated with Gaskell & Chambers, who have in their possession an old cost book of his in which is a list of the moulds, etc., of John Carruthers Crane, Bewdley, purchased by him in 1838.

5347. Yates & Birch, Birmingham: c.1800. Succeeded by James Yates.

Touch Marks

The development of the shape of touches is a valuable clue to dating, even if the maker is not identified, and so it is well worth while to inspect and analyse the shapes, types and designs in fashion from 1600 to 1850. It must be clearly understood that observations are in broad generalities, particularly after c.1660, and that it is trends and tendencies which are mentioned. The older styles continued to be struck concurrently with the new, and a piece was not likely to have been made in the year of striking. The maker probably continued to use his original iron, or a replica, all his working life—say from age twenty-five to sixty-five—forty years; and that is above average. I have pieces as widely

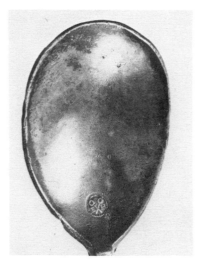

Fig. 8
Spoon-makers' marks. (a) Early, no border, c. 1450. (b) Later, beaded border, c. 1600.

varied in date style, bearing the same touch. Once again, develop your judgment of style, do not rely solely on mark identification; and do not ascribe the date of striking, *ipso facto*, to all pieces with that touch.

The early marks—pre-1620—are so rare that we need not pursue them. If you come across a piece of this date it will speak for itself. We can deal briefly with spoon marks, which, of course, must be small, to fit snugly in the lee of the bowl. A few early marks have no border, and may depict a fleur-de-lys, or a pewterer's hammer, etc., but in the sixteenth century and up to c.1660 they are nearly all small ($\frac{7}{32}''$ to $\frac{1}{4}''$), very well made beaded circles, with two initials and perhaps a key, sword or pellet. In the early half of the seventeenth century the many marks to be found on the handles of flagons are all small neatly beaded circles of $\frac{3}{8}''$ to $\frac{1}{2}''$, with two bold initials with a star, pellet or other simple decoration. They rarely bear a date. I have seen one with 1616, but that is exceptional. Towards 1650 emblems join the initials increasingly, and the diameters are enlarged up to $\frac{5}{8}''$. This question of size is important—beware the piece whose mark is out of step in size! Ninety-five per cent of the marks struck from 1650 to 1660 are circular, but an entirely new fashion in touches now appears—a large oval 1" high, with soft palm leaves within the sides emanating from the lower banner, and supporting the top banner, which bear Christian name and surname respectively.

Fig. 9
An early *dated* touch mark, on a 1" rim bowl-dish of c. 1650. Diameter $15\frac{3}{4}''$. (See Fig. 29.)

In 1666 a spate of small marks suddenly appears on the touch-plate, but this is no new fashion; it is the troubled spoon-makers mentioned earlier who were all ordered to strike new, dated touches.

One still finds occasional small circle touches, with dwindling frequency. For instance L.T.P. 911, struck as late as c.1745, could have been struck ninety years previously to judge by style; but touch-plate No. 4, on which it was struck, was not in existence in 1655.

By 1670 the circular touches are larger, and there are a few more 'ovals and palm leaves', to be followed immediately by a flood of the latter. Also, a very large number of pewterers struck their touches—200 in ten years—despite the danger signals of the economy in our 'golden age'.

By 1673 inclusion of the names was the rule rather than the exception. Pewterers had realised that here was quite good advertising space, and as far as the regulations went, they got away with it. Only 'quite good' advertising space, because the touch marks were shyly tucked away on the backs of plates and dishes, on the rim. You would have been unpopular with your friends if you had declared 'By Charles, this is a fine pewter dish—who made it?' and simultaneously turned it and its contents over, in order to scan the touch. From the earliest examples we have, until the arrival of the early eighteenth-century 'single-reed' type of plate and dish, the touch had almost invariably been on the back of the rim. From c.1700 the touch, for no apparent reason, is always on the back of the well. Perhaps it is because touches are bigger, and need a flat area for even striking. From 1670 to 1694 most of the freshly-struck marks are upright oval, $\frac{7}{8}''$ to $1''$, with

Fig. 10
Drawings of typical forms of touch marks. (a) 'Large oval', c. 1660–c. 1670. (b) 'Upright oval', c. 1670–c. 1694. (c) 'Three-sided dome top', c. 1720+. (d) 'Banners', c. 1735+.

palm leaves at the sides very often supporting the design. Addresses, too, were occasionally added. (Remember that so far we are speaking only of touch marks, not subsidiary marks.) One very interesting mark, L.T.P. 420, dated 1685, includes a dome-lid tankard in the design. This appears not to be the earliest type even, to judge by the rather heavy base. Without the evidence of the date in the touch, one would have dated such a tankard as ten to fifteen years later.

By 1692 the palms tend to dwindle, and the outline often becomes waisted. There are still a few beaded-circle touches, usually $\frac{1}{2}''$ diameter. The designs sometimes show the maker's wares—a baluster, a still and so on. For the greater part, however, the emblems are punning (sometimes extravagantly so) on the surname or addresses, or else are well-executed 'important-looking' designs. By 1705 nearly all new touches were large and more elaborate.

Touches remained similar in appearance until by 1720 a new outline appeared. This was three sides of a square, the top being curved into a shallow dome. The sides of this design are no longer filled with delicate palms, but with strong fluted pillars. It must again be emphasised that new touches of the various types were being struck each year so that types overlapped.

By 1735 the touches had an air of grim starkness, and were less well designed. While the pillared touch was popular, a more or less circular touch of 1″ diameter was even more favoured. There are now prominent banners top and bottom for the name of the maker.

1760 sees the start of other shapes, and in 1780 two or three shields were used. By 1790 some simple ovals appear, some upwards, some across, and smaller than previously. Only twenty new touches were struck between 1820 and 1825, nearly all flat ovals. Pewterers were no longer members of the Company and/or did not bother to strike their touches.

One of the interesting new facts raised is that very, very few baluster measure maker's marks are recorded on the touch-plates, and Cotterell, too, recorded a lower than average proportion.

Secondary Marks

In addition to the maker's mark, starting from the sixteenth century, pieces are found bearing other marks, some of which are easy to accept, e.g. initials, as being those of an owner. But there are other types of mark which will either help or confuse you. By name these are:

Class A 'Hall marks'
Rose and Crown
Crowned X
Labels or cartouches
Class B House marks
hR, AR, WR, WIV, GIV
Verification, or Board of Trade stamps

Class A marks were applied at source, by the pewterer, or by the Hall. Class B were nearly always applied after distribution.

'Hall marks' first appeared about 1630, and the Goldsmiths very soon protested. They are NOT *HALL* marks, since they were not applied by, nor relating to, the Hall—they are simply extra marks of each pewterer, to foster the illusion that pieces might be silver.

60

They bear no date or town mark nor a 'registered' maker's mark. They are the choice of the individual pewterers, and each cartouche is his own design. That is why they are always written in inverted commas. The Goldsmiths were worried and complained repeatedly, without getting any satisfaction. The 'hall marks' of each pewterer are sufficiently distinctive to enable a piece to be identified by only a small part of one mark, if it is clear, and it has been clearly and accurately recorded.

The general fashion of styling varied a little—perhaps by so little that one cannot describe the difference—yet a little experience gives the 'feel' of the period of striking, so if you see a set of 'hall marks' you can judge the earliest possible date of striking. There are usually four, until the nineteenth century, when occasionally three or five appear. Apart from a brief period, they are always struck in a line. At first very small and dainty, they were larger by about 1680, and often an old set of small irons was discarded for a set of bigger cartouches and shields which now had serrations round the border. By 1675 they were in very common use—still on the front of the rims of plates and dishes, and on flat-lid flagons and tankards, but on no other wares. Occasionally

Fig. 11
'Hall marks'. (a) Early, small, c. 1640–c. 1670. (b) Later, bold, c. 1665+. (c) Struck at random (usually on the lids and drums of flat-lid flagons and tankards). c. 1690–c. 1700.

61

flat-lid tankards and flagons around 1700 bear 'hall marks' apparently struck at random, both on the lid and drum. About 1700 they suddenly took shame and hid on the back of the well of both plates and dishes, but are to be seen on the top of the drum of flagons and tankards, both lidded and lidless.

Occasionally the 'hall marks' of one pewterer appear on a piece touched by another. There had probably been a change of ownership of the firm and the 'hall mark' irons handed on, either within the family or by purchase of the business. Very occasionally two different makers' 'hall marks' appear on one piece: in these cases it was probably subcontracted out, and both showed their credits.

'Hall marks' of all periods consist of heraldic devices, black letters, and maker's initial or initials. True silver hall marks are sometimes closely copied. 'Hall marks' are little more than a means of identification, and much less informative than the touches. In the mid-nineteenth century they are very often two devices repeated.

Perhaps this is letting a large cat out of the bag, but here is some useful experience I have not seen printed elsewhere. 'Hall marks' do NOT appear on genuine salts, flagons pre-1650, candlesticks of the seventeenth century, baluster measures, lidless tankards and dome-lids before 1700, and very seldom on porringers. That is an almost inflexible rule, and any pieces mentioned so adorned should be greeted with the greatest discourtesy and suspicion. There *may* be some; the fact that I have not seen or heard of them does not mean that they could not exist. Experience has taught me the danger of making positive and exclusive statements!

'Hall marks' arrived in Scotland and Ireland very much later—well into the eighteenth century. In Scotland the Thistle was usually incorporated, and in Ireland the Harp.

Rose and Crown

How far back goes the use of the Rose and Crown by the Company is not recorded, but it was in use in 1566, and in fact pewterers were warned then not to add their initials to it in making their touch. It is obvious that it was, for some reason, a mark of quality. (Tin ingots were similarly stamped.) It implies that, where warranted, the pewterer applied the Rose and Crown. I am fortunate to own a plate of at latest (probably earlier) Tudor period which bears this Rose and Crown struck on the face of the rim. The fact that there is no touch mark could incline one to believe it to be of fifteenth-century manufacturer, and would therefore place the

Rose and Crown in that century—seventy years earlier than the record mentioned above. The next relevant reference is a hundred years later, when this mark was used not merely for quality, but on pieces for export—again presumably struck by the maker. It was still regarded as a symbol of quality, for by c.1680 pewterers struck a Rose and Crown as an additional mark on their plates and dishes, alongside their touches, of the same size and principles of design. Each pewterer used his own design of Rose and Crown, and so with diligence and adequate recording, theoretically, one can trace a maker by his Rose and Crown alone. Before long

Fig. 12
Rose and Crown used as a subsidiary mark, alongside the touch mark. c. 1690. (Appears only on plates and dishes.)

several were defying the Company and incorporating a Rose and Crown in the design of their touches. This persisted well into the nineteenth century. In Scotland it was enacted that pewterers should use the mark of a Thistle, and the Deacons mark, and the quality of pieces bearing them was to be equivalent to that marked in England with the Rose and Crown. Is it not strange that so few pieces have come down to us marked in this way? Mid- and early seventeenth-century dishes seldom if ever bear it, and our standby examples of the early part of the century—flagons—never. Before leaving the Rose and Crown, bear in mind the difference from its use as a touch in the Low Countries. There the petals of the rose are filled with fine radius lines, and/or the crown bears, or contains within its outline, the makers' initials (see Fig. 1).

The Crowned X

Although this mark appears in the base or on the lip of so many Victorian tankards, its history goes back to the end of the seventeenth century, but its use earlier still can be inferred by the injunction to stamp the X only on 'Extraordinary ware called Hard Mettle'. Certainly, pieces of this period that bear it *are* hard, and may have a higher tin content than normal. Cotterell points out that basically it may be a use of Roman numerals to denote a 10 : 1 proportion of tin, which is known to have been laid down on the Continent as the standard for best quality use. This would reduce the lead content to less than half the old standard for hollow-ware. The relative proportions of lead : tin in 'Extraordinary' and 'Standard' are thus 1 :10 and 2.3 : 10. Of course, it may be much simpler—X may be short for 'Extraordinary Ware called Hard Mettle'! Its use became widespread, and not always qualified by the quality of metal. Have we not noted that control was lost? Certainly one can find the crowned X throughout the eighteenth and nineteenth centuries, on various types of ware for no noticeably justifiable reason.

Labels or Cartouches

We have seen the pewterers' sudden desire for self-advertising, and some of the opportunities taken in the latter seventeenth century. There was no holding those publicity-minded boys. In addition to touch mark, 'hall marks', Rose and Crown, questionable use of X, the enterprising pewterer further attacked the market with cartouches bearing at first 'London', then 'Made in London', for prestige purposes. Provincial pewterers followed their example and did the same—'London' included!—and it transpired that nothing short of an Act of Parliament could stop them. The London makers stepped up the campaign by adding their address, perhaps in even another cartouche, and several added others containing such more or less meaningless jargon as 'Superfine Hard Metal'. Presumably their 'commercials' paid off, for such labels persisted throughout the eighteenth century.

House Marks

This is the title given to marks very similar in appearance to touch marks, but which denote the ownership of the piece—and inn or abbey, etc. They are struck with an iron just as a touch mark. (Engraved ownership does not come under this category.) It is fair to include corporation ownership such as the Great Yarmouth coat of arms to be seen on eighteenth-century plates and balusters. In Peterborough Museum there is a much earlier example, a sixteenth-century plate bearing a Ram on the rim. This is a rebus for the abbey at Ramsey, near which the plate was excavated. Usually these rather rare marks appear on baluster measures. Note that true house marks are struck from irons as well made as touch marks. Beware any that are crudely made! These will be fakes, and there were many on the market, before the Second World War. And they are still in existence, of course, prized and changing hands, both at home and abroad. (See chapter on 'Fakes'.)

64

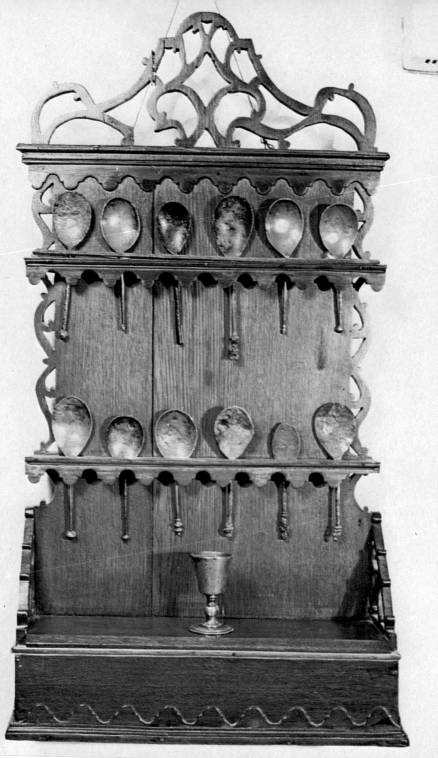

Plate 5 15th c. spoons. Writhen Ball; Latten knop; Stump; Horned Headdress; Latten knop; Golf Ball; Peachstone Ball; Diamond; Baluster; Crown; Genuine but uncertain spoon; Lion Guardant. (Stump only bears a mark)

Plate 6 (a) 1605 flagon. These seldom bear a mark. 1605–c. 1620. Ht. to lip, $8\frac{3}{8}''$.
(b) Bun lid flagon. Mark, "C.B." on handle. c. 1630. Ht. to lip, $9\frac{1}{2}''$

Fig. 13
House marks, usually of an inn (the Saracen's Head). Besides being an emblem, easily recognised by the illiterate, they were an aid to security. The owner's initials often appear. In the sixteenth and early seventeenth centuries they were often struck inside the neck of the measure as well as on the lid. (This 'wedge' *must* have carried a proper thumbpiece.) Early seventeenth century. Mark 'I'. Height to lip $5\frac{3}{4}''$.

The house mark almost invariably contains the two, or three, initials of the innkeeper, and the emblem of the house. It is usually struck three or five times on the lid of the baluster. Earlier it was also struck once inside the neck. Besides this mark, the measure will very often carry the same initials of the landlord stamped on the lid and on the handle for thorough security. A very rare form, so rare as to be almost negligible, is the house mark branded on the inside of the base. But after, say, 1650, where used, they are always prominently placed.

Fig. 14
Very occasionally the house mark was branded on the base inside. The Bull, on another fine 'wedge'. Early seventeenth century. Mark 'D.B.' with ?. Height to lip $7\frac{3}{4}''$. (See Fig. 31a.)

hR, WR, AR, WIV, GIV (each usually with a crown)
The first of these caused much misdating at first. It was assumed by Massé that this stood for Henricus Rex—Henry VIII. Very reasonable, seeing that it occurred on battered, early-looking buds and hammerheads, and that some other balusters and tankards bore

65

AR and WR. It looked as if these marks, struck on the lip or lid of balusters, referred to the reigning monarch. Thus some buds we now know to be of c. 1700 were dated as mid-sixteenth century. Then it was suggested, after a logical redating of styles of measure, that they referred to the standards of measure enacted in those periods. They need inspecting separately. The meaning of hR (small 'h') is still not solved, and there are very few even remotely feasible suggestions. We can reject the small 'h' as standing for Henry. Michaelis, *A.P.B.I.*, suggests that in view of the interest in correct measure of 'muggs' by the Clerk of the Market of the Queen's Household in 1708, it stands for 'household Royal' (or more likely 'Reginae'). It seems far-fetched to think that the many different makers whose measures bear this mark all belonged to the royal household, but this official's position included jurisdiction over an area where at least stone pots were made, so there may have been pewterers too under his control, or even markets where the 'muggs' were sold. I am not convinced, but certainly have nothing better to suggest. At present it just means that you have something of an enigma which looks very well on balusters. WR was used on balusters and tankards, and is undoubtedly the seal of a verification check of capacity, conforming with the Act of 1688, standardising the Ale and Wine measures, which explains why WR appears on some of either capacity range through the eighteenth century. (I have a Society flagon of 1819 bearing WR.)

AR (Acts of 1704 and 1707) was no doubt used for similar reasons, but somehow use of these later irons was discontinued after the death of Anne, in favour of WR. One feels that these three check stamps were probably applied after distribution, as one would not expect the pewterer to have sales big enough to justify keeping large stocks. On the other hand, GIV and WIV were stamped by the Excise authorities as proof of Imperial capacity, but as the tankards were now produced by larger firms it would be far more convenient for Excise officials to visit the workshops to check and stamp stock.

Owners' initials are easily accepted, bearing in mind that 'I' often stood for 'J' to the end of the eighteenth century ('J' starts to appear c. 1720). Broadly speaking, until the nineteenth century only two initials were used—more would indicate two persons. Given the triangle of initials, e.g. $\frac{I}{IM}$, the top initial is that of the surname, the others are the initials of husband and wife. In this case it could have been 'Jones, James and Mary'. Occasionally

Fig. 15

Verification stamps. (a) hR. Possibly c. 1600–c. 1720, certainly c. 1670–c. 1700. (b) WR. Conforming to Act of William III. Used c. 1680–1826. (c) AR. Apparently confined to Anne's reign. (d) WIV, GIV. Confined to their reigns respectively.

67

initials may be found surmounted by crowns. Think nothing of it, for it has no known significance.

Verification Marks

We will only touch on capacity standards by noting that in 1826 Imperial Standard was implemented. Quite frequently one sees measures and tankards bearing a portcullis, 1826 above stamped on the rim. The action of stamping was probably not confined to 1826, but confirmed that the contents conformed with the new standard. This stamp was probably, within a year or two, followed by the use of GIV (whose reign was 1820–30) and this in turn by WIV (1830–7). We have assumed that workshop-produced goods

Fig. 16
The first Imperial Measure verification stamp, c. 1826–c. 1830. (A portcullis, with 1826.) Actual width ¼″.

were verified at source, but subsequently measures sometimes bear a sequence and multiplicity of verification marks. One wonders if certain inn-keepers were regarded with grave suspicion, and the authorities chivvied and hounded them constantly. I have seen about fifteen verification stamps on a single tankard. Note that it is not only measures which are verified, but also tankards in general use. After the omnibus GIV and WIV stamps, County and Borough stamps were used, usually being a simplified reproduction of their heraldic arms. Nowhere have these been published to date of writing, but there are hopes that they will be. In the meantime the local library or museum could probably help. These stamps of emblems or arms often have a letter or numeral, denoting the year, coded. In 1879 central control and standardisation was more or less imposed, and the stamps were the monarch's initials crowned, with a number and, in some districts, a letter,

Fig. 17
Borough and County marks were used for verification stamps c. 1830–1878. Norwich VR.

e.g. $\frac{VR}{159}$. The number was allocated to a county or borough, and the letter denoted a year. The numbers were issued one by one, as applied for—NOT usually as a group. Thus two consecutive numbers would probably be from widely different localities. Some fifty-two in England and Wales, sixteen in Scotland, and one in Ireland opted out of the scheme, and kept to their own ideas.

If the need for a particular stamp number (district) declined, it was taken out of use and the number held in suspense. Before being reissued elsewhere it was offered again to its old locality. Therefore if one is keen to trace the places where a tankard was used in the bar, it is best to use the *oldest* book available, which will be one of the Weights and Measures Inspector's handbooks. There is a danger in publishing a list in that it is possible that some district numbers may have been reallocated. However, I am very kindly allowed to publish the list supplied by the Board which appears in the 1950 edition of the Handbook issued by the Institute of Weights and Measures Administration. This will be of inestimable interest to collectors of the latter period pieces, and will facilitate research into provenances, distribution, etc. The local Weights and Measures office will usually be pleased to try to help. It would probably be both helpful and tactful to submit a rubbing or drawing by post, to allow the very busy Inspectors to deal with your enquiry at their 'leisure'. I for one am very indebted to my county Inspector for his co-operation, and I have always found the Board's officials keenly interested in all the historical aspects of their work.

OFFICIAL STAMP NUMBERS
(Weights and Measures)
(Revised to 31st December, 1950)

Where there is no entry in the column headed 'District' it implies that the Stamp Number either is unappropriated or is not now in use by the Local Authority to whom it was originally allotted.

No. of Stamp	District	No. of Stamp	District
1	Board of Trade	49–50	Westmorland, County
2	London, City	51	Huddersfield, Borough
3	Edinburgh, City	52–53	Lanark, County
4	London County Council	54–57	Cheshire, County
5	Manchester, City	58	
6	Birmingham, City	59–62	Glasgow, City
7	Nottingham, City	63	Renfrew, County
8	Bedford, County and Dunstable Borough	64	Sunderland, Borough
		65	Wolverhampton
9–12	Cornwall, County	66–67	London, County
13	London County Council	68	Plymouth, City
14	Bradford, City	69	Perth, County
15	Banff, County	70	Wigan, Borough
16	Renfrew, Burgh	71	Newcastle-on-Tyne
17	Renfrew, County	72	Paisley, Burgh
18–19	Derby, County	73	Perth, City
20	Margate, Borough	74	Poole, Borough
21	London County Council	75	Plymouth, City
22	Angus, County	76	Salford, City
23	Coatbridge, Burgh	77	West Sussex, County
24	Lanark, County	78	
25	Bradford, City	79	Wiltshire, County
26	Stafford, County	80	Scarborough, Borough
27	Maidstone, Borough	81	
28	London, County	82	Stafford, Borough
29	Middlesex, County	83	Stalybridge, Borough
30	London, County	84	Glasgow, City
31	Middlesex, County	85	King's Lynn, Borough
32–33	Stafford, County	86	Birkenhead, County Borough
34–36	Glasgow, City		
37	Sheffield, City	87	
38	Bath, City	88–94	Lancashire, County
39	Bedford, Borough	95–96	London, City
40	Plymouth, City	97	Lancashire, County
41	Kingston-upon-Hull	98	
42	Canterbury, City	99–102	Lancashire, County
43–48	Cumberland, County	103	

No. of Stamp	District	No. of Stamp	District
104	Lancashire, County	179	
105–107	Bradford, City	180	
108–109	Lancashire, County	181	
110	Leicester, City	182–189	Essex, County
111	Montrose, Burgh	190	Sutherland, County
112–113	Norwich, City	191	Roxburghshire, County
114–118	Buckingham, County	192	
119		193	
120	Oxford, City	194–195	Suffolk, West
121	Chester, City	196	
122	Dumfries, County and Borough	197	Brighton, Borough
		198	Boston (Lincs.)
123	Brechin, Burgh	199	Bolton
124	Weymouth and Melcombe Regis	200	Blackburn
		201–202	Ayr, County
125	Isle of Ely	203	
126		204	Ayr, County
127–128	Birkenhead	205–206	
129	Lincoln, County	207	Colchester, Borough
130	Kesteven, County	208	Coventry, City
131–134	West Bromwich	209	Fife, County
135–136	Wiltshire, County	210	
137–140		211–212	East Sussex
141–143	Derby, Borough	213	Greenock, Borough
144	Grantham, Borough	214	Hereford, County
145	Derby, Borough	215–216	Sussex, East, County
146	Aberdeen, City	217	Hove, Borough
147	Liverpool, City	218	Eastbourne, County Borough
148	Darlington, County Borough	219	East Ham, County Borough
149	Plymouth, City	220	High Wycombe
150	Northallerton	221–222	East Ham, County Borough
151–152	Yorkshire, N. Riding		
153	Plymouth, City	223	Ipswich, County Borough
154–155	Durham, County		
156	Northampton	224	Kidderminster, County Borough
157	Durham, County		
158	Northampton, Borough	225	
159–160		226	Kilmarnock, Burgh
161–163	Dorset, County	227	
164–168		228	East Ham, County Borough
169–170	Durham, County		
171–173	West Sussex, County	229	Kirkcudbright, County
174	Leeds, City	230–231	
175		232	Louth, Borough
176–178	Wolverhampton, County Borough	233–234	

71

No. of Stamp	District	No. of Stamp	District
235	Newark, Borough	278	Bootle, County Borough
236	Newport (Mon), County Borough	279	Barrow-in-Furness
		280	Glamorgan, County
237–238		281	Southport, County Borough
239	London, County		
240	Moray and Nairn, Counties	282	Glamorgan, County
		283	Anglesey, County
241	Walsall, County Borough	284–289	Glamorgan, County
242		290	Clitheroe, Borough
243	Walsall, County Borough	291	Berkshire, County
244	Tynemouth, County Borough	292–293	
		294	Irvine, Royal Burgh
245	Walsall, County Borough	295	
		296	Swansea, County Borough
246	Great Yarmouth, County Borough	297	Lincoln, City
247		298–320	West Riding, Yorkshire
248	West Lothian, County	321	Radnor, County
249	Berkshire, County	322	Hartlepool, Borough
250	Hastings, County Borough	323	Gateshead, County Borough
251	Selkirkshire	324	South Shields, County Borough
252	Cardiff, City		
253	Warwick, County	325	Dumbarton, County
254	Reading	326	Batley, Borough
255	Penzance, Borough	327	Luton, Borough
256	Rothesay, Burgh	328	Newcastle-under-Lyme, Borough
257	Bute, County		
258	Inverness, County	329	Stirling, County
259	Brecon, County	330	Soke of Peterborough, County
260–261			
262	Berwick-on-Tweed, Borough	331	Crewe, Borough
		332–333	East Riding, Yorkshire
263		334–340	
264	Worcester, City	341–343	Kent, County
265	Southampton	344–346	London, County
266	Huntingdon, County	347–348	Kent, County
267–269	Suffolk, County East	349–351	London, County
270	West Bromwich, County Borough	352–357	Kent, County
		358	Royal Tunbridge Wells, Borough
271	Suffolk, East		
272	Suffolk, West	359	
273	West Riding, Yorkshire	360	London, County
274		361	Clackmannan, County
275–276	Glasgow, City	362	Stirling, County and Burgh
277	Warrington, County Borough		
		363	Selkirkshire, County

No. of Stamp	District	No. of Stamp	District
364–365		430	
366	Hereford, City	431–432	Kent, County
367	Oxford, County	433	
368		434	
369	St. Albans, City	435	Warwick, Borough
370–376	Smethwick, County Borough	436	
377		437	Stockton-on-Tees, Borough
378	Somerset, County	438	Dunfermline, Burgh
379	Burton-upon-Trent, County Borough	439	Doncaster, County Borough
380		440–441	
381	Glossop, Borough	442–444	Flint, County
382	Gloucester, City	445–446	
383	Merioneth, County	447	Portsmouth, City
384	Stockport, County Borough	448	Edinburgh, City
385	Surrey, County	449–452	
386	London, County	453	Berks, County
387–389	Surrey, County	454–455	Wiltshire, County
390–391		456	
392	Carlisle, City	457	Southend-on-Sea, County Borough
393	Accrington, Borough	458–459	Carmarthen, County
394–399	Hertford, County	460	
400–401	Leicester, County	461	Glamorgan, County
402		462–464	Wilts, County
403–405	Leicester, County	465	Caithness-shire and Wick, Burgh
406–407			
408–411	Gloucester, County	466–467	Rotherham, County Borough
412			
413		468	Ayr, Burgh
414–415	Chesterfield, Borough	469	Rotherham, County Borough
416	Lancaster, City		
417	Ashton-under-Lyne, Borough	470	Morley, Borough
		471	Hamilton, Burgh
418	Chesterfield, Borough	472	Bacup, Borough
419–421	Pembrokeshire, County	473	St. Helens, County Borough
422	Forfar, Burgh		
423	Aberdeen, County (Kincardine)	474–476	Warwick, County
		477	Burnley, County Borough
424	Oldham, County Borough		
		478	St. Helens, County Borough
425			
426	Berwick, County	479	Midlothian, County
427	Cambridge, Borough	480	West Ham, County Borough
428	Montgomery, County		
429		481	Neath, Borough

No. of Stamp	District	No. of Stamp	District
482	Nottingham, County	540	Wallasey, County Borough
483	Croydon, County Borough	541–544	West Ham, County Borough
484	Hertford, County	545	York, City
485	Stoke-on-Trent, City	546	Kirkcaldy, Burgh
486	Peterborough, City	547	Oxford, County
487	Dundee, City	548	
488	Arbroath, Burgh	549	Guildford, Borough
489	Warwick, County	550–552	Norfolk, County
490	Bristol, City	553	Blackpool, County Borough
491	Macclesfield, Borough		
492	Cheshire, County	544	East Lothian, County
493		555	Airdrie, Burgh
494	Nottingham, County	556	Shrewsbury, Borough
495	Preston, County Borough	557	Norfolk, County
496	Middlesbrough, County Borough	558	Lincoln, County (Holland)
497–499	Denbigh, County	559	Hampshire, County
500	Rochdale, County Borough	560	Barnsley, Borough
501–506	Northumberland, County	561	
507	Maidstone, Borough	562	Bury, County Borough
508–509	Northumberland, County	563	Monmouth, County
510	Monmouth, County	564	Inverness, Burgh
511	Hampshire, County	565	Cambridge, County
512–513	Monmouth, County	566	Argyll, County
514	Kendal, Borough	567	Salop, County and Wenlock Borough
515–517	Caernarvon, County		
518–519	Norfolk, County	568	Ross and Cromarty, County
520	Isle of Wight, County		
521	Dudley, County Borough	569	Dover, Borough
522–524	London, County	570–571	Cardigan, County
525–526	Northampton, County	572	Reigate, Borough
527	East Riding (Yorks), County	573	Folkestone, Borough
		574	Wilts, County
528	Lincoln, County	575–577	Somerset, County
529	(Lindsey) County	578	Merthyr Tydfil, County Borough
530	Warwickshire, County		
531	Rutland, County	579	Nottinghamshire, County
532	Grimsby, County Borough	580	Gravesend, Borough
		581	
533	Lincolnshire	582	Bournemouth, County Borough
534	Worcester, County		
535–536	Aberdeen, County	583	Devon, County
537	Worcester, County	584	Exeter City (Devon, County)
538–539	Barnsley, County Borough		
		585–588	Devon, County

No. of Stamp	District	No. of Stamp	District
589	Middlesex, County	619	Dumbarton, Burgh
590	Wakefield, City	620	Carmarthen, Borough
591	Halifax, County Borough	621	
		622	Durham, County
592	Dewsbury, County Borough	623	Merthyr Tydfil, County Borough
593	Wigtown, County	624	East Riding (Yorks) County
594	Stirling, County		
595		625	Northampton, County Borough
596	Devon, County		
597	West Hartlepool, County Borough	626–627	Dundee, City
		628–630	Cheshire, County
598	Hyde, Borough	631	Devon, County
599–609	Essex, County	632	Devon, County (Okehampton)
610	Worcestershire, County		
611		633	Cheshire, County
612		634–635	Coventry, City
613		636	Northampton, County
614		637	Newport, County Borough
615–618	Cardiff City		

PART II

5. Romano-British, Early Medieval—to End Sixteenth Century

The Romans came to England, not primarily to make pewter, as you and I might think, but for other economic and prestige purposes. They knew of our tin, and as well as using it for consumer and military products they copied silver in it. Tin, however, is very harsh, and liable to decompose. Somehow—perhaps just to make up substance—some lead was added, and the alloy no doubt proved easier to work, and certainly better in use. The well-known hoard found at Appleshaw in the middle of the last century included items containing varying proportions of lead to tin. The 'Fish' dish was 99·2 per cent tin, other pieces varied down to 90 per cent; other pieces of Romano-British in the country go down to only 47 per cent tin! Soon they appear to have settled at 62–80 per cent—which latter was the medieval standard. The hoards have been discovered principally in East Anglia, and the pewter was virtually confined to the civil zone south-east of the Fosse Way. East Anglia was subject to fierce, violent raids in the fifth century A.D. Warning was short—a hole rapidly dug—the pewter hidden—the raid successful, and the owners were left in no fit state to recover their fine dishes. There the dishes remained until they reappear during deep ploughing, and building of many airfields. So East Anglia has provided perhaps disproportionate evidence of distribution due to (a) fifth-century raids, (b) the Second World War and (c) deep ploughing.

Manufacture has been proved at sites in the Bath area, notably at Camerton. (See *Excavations at Camerton* by W. J. Wedlake.) Tin was brought from Cornwall, and lead, stone (for moulds) and coal were available close by in the Mendips. The products are invariably well made, some of superb craftmanship. For design, technique, decoration and enigma, the octagonal flanged bowl with Chi Rho and other presumably Christian decoration, now at the Museum of Archaeology and Ethnology, Cambridge, is perhaps the finest piece of pewter in the country of any period. This is

Fig. 18
Romano-British octagonal flange bowl. Possibly one of the finest and most interesting pieces of pewter in existence. The upper face of the flange bears many early Christian symbols—not yet fully elucidated. Found at Ely. c. 300 A.D. Height 4″.

(*Dept. of Archaeology and Ethnology, Cambridge*)

Plate 7 (a) John Emes flagon (solely his speciality). (Cott. O.P. 3092) c.1690. Ht. to lip, 8″. Very rare. (b) Beefeater flagon. Touch branded in base, "hall marks" on lid. Maker, T. Lupton, c.1660. Ht. to lip, 8⅜″.

Plate 8 (a) Flat lid flagon, two bands, deep lid. Mark, "I.F.". c.1685. Ht. to lip, 6¾″
 (b) Flat lid flagon, plain, shallow lid. Mark "W.W." (Cott. O.P. 6028). c.1695. Ht
to lip, 8⅛″ ("Hall marks" scattered on lid)

probably true judged on any two of these points, excepting the decoration. The decoration consists of Chi Rho, the Greek symbol of Christianity, and other emblems of the faith not yet elucidated. It is very clear, and obviously of great potential interest. Items made in the period range from chargers of 32″ diameter down to saucers of 5″ diameter; little cups for sweets, sauce or spices; bowls, plain or with circular or octagonal flanges; ewers; jugs; caskets (perhaps for cremation remains); even to a candlestick.

Some of the dishes bear a decorated panel in the centre, based on geometric and what can best be described as squirl designs. The rim types vary considerably in detail, and some are very similar indeed to the single-reed plates of the eighteenth century. They

Fig. 19
A fine and typical example of a Romano-British plate. The concentric grooves were almost invariably present, underneath to take the weight of a standing ewer. c. 350 A.D. Diameter 15¼″.

were cast in stone moulds and trimmed on a lathe, but they were not hammered for strength. One certain method of identification is that *all* plates and dishes carry a support ring underneath, of approximately half the diameter of the well. Therefore they were used to carry weight—probably a jug or ewer, for knife marks are never to be seen on them. Also the great majority have three or four concentric grooves on the top of the plate, from a quarter to a third of the diameter of the well. No piece has been found with a maker's mark. ('Romano British Plates and Dishes', Christopher A. Peal. *Proceedings of the Cambridge Antiquarian Society*, Vol. LX, 1967.)

Indeed the Romans discovered pewter—and several museums

have pieces on display. They very soon achieved a high standard of workmanship and established desirable alloy standards. I wonder if the early medieval pewterers were aware of their predecessors' skill. When the Romans left Britain, trade and communications collapsed, so pewter manufacture lapsed as a trade.

A few Anglo-Saxon brooches of base alloys are the sole existing representatives of a millennium ago. We have no evidence of domestic ware until ten years after the Norman invasion. Perhaps the Normans are a significant link, for it is not known how or where or whence pewter was resurrected.

1076 is an oft-quoted date, for it is the first historical record of pewter, in which it is referred to as 'tin'. This confusion of term persists in German and French, for their nouns are 'Zinn' and 'étain'. 'Étain' mutated to English would be 'stin' and the common derivation (Latin 'Stannum') is obvious. 'Pewter' is a corruption of Old French 'peutre'. One wonders if the English 'tryffle' or trifle is derived from 'peu-tre', perhaps 'little' or 'lesser tin'.

The record of 1076 refers to 'tin' being allowed for Church vessels, and Michaelis comments that there are other written evidences of church pewter in the eleventh century. We now meet the first examples of early medieval pewter—the sepulchral chalices, and occasionally patens, which were buried with monks in their coffins in the twelfth and thirteenth centuries. This was specifically requested by the Bishop of Winchester in 1229.

There are one or two spoons of very early date, probably of the thirteenth century. There are two very fine and well-preserved ecclesiastical cruets which are most probably of the thirteenth to fourteenth century, at Ludlow, and Weoley (Birmingham). Their similarity is quite remarkable. Both were found in the moats of their respective castles. In 1348, another memorable date, the first recorded representations were made, which culminated in the charter being granted for the founding of the Pewterers Company in 1473. Obviously for much more than this hundred years the makers had been getting together for their self-protection, and determining standards of workmanship and alloy.

Before leaving this period considerable attention can be paid to a remarkable bowl, the condition of which guarantees it to be completely genuine, but which appears to be of Anglo-Saxon origin. One deduces its function as being a font bowl, by its depth and by the striations down the outside of the bowl on the sides— caused by repeatedly rubbing the coarse stone of the font. The base is comparatively unworn. The rim bears four nicks, not quite

Fig. 20
Possibly one of the earliest entire pieces of purely British pewter. A font bowl bearing late Anglo-Saxon decoration carried out in their technique. Probably a rural perpetuation, but *may* be a missing link. Probably twelfth to fourteenth century. No mark. Diameter 15½".

regular in spacing, possibly for positive positioning. Unfortunately it bears no maker's mark. Its very great interest is locked up in the decoration on the rim, which is almost pure Anglo-Saxon type, and is illustrated. Despite considerable research, no near parallel can be found. It is even more significant that the manner of execution is that used on Anglo-Saxon silver (and Roman pewter)—the chip carving method—the lines are punched with a small chisel. The background in the decorated panels is stippled

83

in parts. Furthermore the well of the bowl is traced out for additional decoration, in the Anglo-Saxon fashion and manner. Its remarkably fine preservation, with a heavy layer of oxide, is no doubt due to the little use it was subjected to in a font, and is an excellent example of a piece which should NOT be cleaned, much as I would like to do so. It is probable that this is a rural perpetuation of Anglo-Saxon craft, used in the time-honoured system of grafting one culture or religion onto and into the next, and

Fig. 21
Very early flagon, probably fifteenth century, dredged from the Thames. It looks more Continental than English, but there is so little contemporary evidence for comparison. No mark. 6¾″ to lip.
(Mr R. Mundey)

since this turned up in a very remote area it may be twelfth century. I may be out by 300 years. We have nothing to compare. This, and the Chi Rho octagonal bowl at Cambridge, I regard as the most interesting pieces of pewter in existence. This is quite possibly the earliest piece of entire pewter of English manufacture.

If your eyes are open you may come across something equally exciting. Others had seen this before I had the luck. Anyone might see a comparably unique piece somewhere, some time. But it requires the confidence born of experience and knowledge to assess such an item.

In the fifteenth century there were founded not only the Pewterers Companies of London, but also of York, Bristol and Edinburgh. At the same time Guilds of Freemen of associated crafts

Fig. 22
A very early hammerhead baluster. The early forms were based on pottery jugs, and are very slim and primitive. Maker's mark, illegible. House mark, a bell and R.C. In the base, a star, cast. Height to lip 9″.

(Mr R. Mundey)

were established at many centres. Records refer to very many types of ware which are not defined, and of which we have no knowledge. In fact in many cases it is not clear whether the reference is to type or size. Michaelis, *A.P.B.I.*, quotes many instances and gives his references. There are only very few pieces of hollow-ware which probably belong to this period. Three domestic flagons come to mind, in the Norwich and Hitchin Museums, and in the possession of Mr R. Mundey. It is possible that some of the very early slim balusters (e.g. Mr Mundey's) are of the fifteenth century, but it is more likely that they are a little later. The greatest numerical evidence of the period is in spoons, for this is the time when the pewterers' lobbying and pressure appears to have virtually stamped out competition. The spoons are with fig- or pear-shaped bowls, and the thin diamond-shaped stem has now been tapped on top and underneath to form a hexagonal stem, which

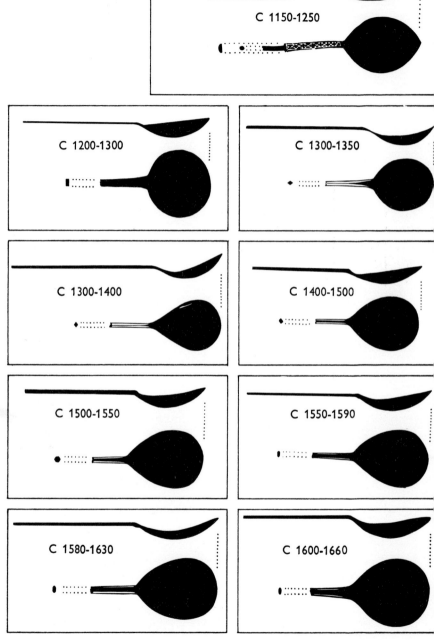

Fig. 23
Silhouettes of spoons, c. 1200–c. 1665.

(*By courtesy of 'The Connoisseur'*)

by the end of this period became rather heavier. (Peal, 'Latten Spoons', *The Connoisseur*, April, July 1970.) The types which run in the fifteenth century include Writhen Ball, Diamond Ball, Acorn, Stump End (which is very different from Slip Top), Horned Headdress, Melon, some Maidenheads, and some earlier Latten Knopped spoons (Plate 5). Do not forget that some types ran well into the next century—the Diamond Knop had a life of some 200 years, and while *some* Acorns are early, most are sixteenth century. Latten was used to simulate the gilt knops on silver spoons. The Horned Headdress is the most commanding of all types of knopped spoon, and has been raped and aped by the fakers. Be very wary!

Fig. 24
One of a hoard of about twenty plates, all bearing the feather mark on the face of the rim. This was the emblem of Prince Arthur, 1486–1502. This plate is one of the very few surviving pieces of copper/tin alloy. No maker's mark. c. 1500. Diameter 10½".

Remember that whereas all items in silver were carefully preserved, old pewter was only good for scrap. And so all existing pewter spoons have been recovered from drains, wells, the ground, river, mud and so on—which helps to explain why pewter is so very much more rare than silver. There are a few spoons in the Victoria and Albert, and Salisbury Museums, and odd ones in many other museums.

Some of the bumpy-bottom plates almost certainly belong to this period. A hoard of some twenty was discovered in 1899 during extensions to Guy's Hospital. They are of copper/tin alloy, exceptionally fine, and each bears an ostrich feather in a long rectangular mark on the front of the rim. This was the emblem of Arthur, Prince of Wales (1486–1502). Specimens of early bumpy-bottom plates probably of that century may be seen at the Department of Archaeology and Ethnology, Cambridge, Weoley Castle,

Guildhall, and Victoria and Albert Museums. Most, however, belong to the sixteenth century or early seventeenth century.

Sixteenth Century

These bumpy-bottom plates are charming, and reflected light throws up their design most attractively. They are, of course, very rare but I was fortunate to find a small dish of this style a few weeks before writing. The rim is rather wide and plain. Two similar are know, both dated 1585, but with punched decoration forming a series of scallops. Specimens may be seen at Hampton Court and the Guildhall Museum.

Fig. 25
A fine bumpy-bottom dish of late Tudor period. Mark illegible. Diameter 14½".

Although there are many records of the doings and mis-doings of pewterers, which we have already touched on, only a few types are described, and our examples of hollow-ware of the period are woefully inadequate. We know that tankards were made. We have none. Stone pots with pewter lids threatened the Company, and were probably similar to foreign examples as in, for instance, Fitzwilliam Museum, Cambridge.

Undoubtedly some of the early baluster measures are of this period (Guildhall, Victoria and Albert Museums, Pewterers Hall), which cannot be later than sixteenth century. Although we have no archaeological evidence, pictures or descriptive evidence to define styles, and we cannot *with certainty* date earlier than the end of the sixteenth century, it is very likely that these were made continuously over a longer period. The very early feel and form of Mr Mundey's examples, and the Guildhall's, suggest that some

Fig. 26
The body of an early half-pint O.E.W.S. baluster. Note the primitive lines with no flaring at the base. This bears a house mark of a Rose and Crown under the base. (The base has been tapped up to reduce contents!) Mark 'F.B.' Height $4\frac{1}{2}''$.

could certainly be up to 100 years older, and maybe more. One of my own, unfortunately without lid or thumbpiece, shows exceptionally early lines—see the pure curve and narrow base. Who is to say that any of these is the first of the type to be made? I feel that balusters, in some form, certainly existed in the fifteenth century, perhaps earlier, and that some existing examples may date to then.

Candlesticks were made with a bell base (and probably other styles as well). One such is illustrated and another similar is to be seen at the Victoria and Albert Museum, but the base of the latter has been mis-reshaped. These are the earliest known. This type is recorded recognisably, as two sizes, and that illustrated is the smaller at '2 pounds the pair', and is $8\frac{1}{2}''$ high. Salts, both Bell and Trencher, were made. None of the latter has been identified, but I found the base of one of the former acting as door stopper in an antique shop. Identical styles were made in Holland, and the goods of the Heemskirk expedition of 1596 were recovered early in this century, and there are several examples from it in the Rijksmuseum, Amsterdam. (Note that the intense cold of 300 years in Nova Zembla did not cause 'Tin Pest' in the pewter! Some pieces are, by contrast, a little corroded.)

Fig. 27
Bell candlestick—a hint of the glories of
Tudor design. No mark. c. 1690. Height
$8\frac{3}{8}''$.

Probably very little pewter was used in the churches in this century, but a most interesting example which may be of this period
is at the Pewterers Hall. This is bulbous, on a thin stem over the
foot. The weakness at this point may account for this being the sole
survivor of the style. It is similar to silver examples, but one should
add that it appears to be similar in style to a flagon depicted in
the well-known touch of E.G. of about 1630, and so it is more
probably early seventeenth century. There is a very fine example
of quite different style of the sixteenth-century period at the
Guildhall Museum. Curved-sided, two-ear porringers were made,
but are exceedingly rare. The sixteenth century saw the heyday of
pewter spoons. Many types of knop are known of this period, including Gauntlet Seal, Lion Sejant, 'Peachstone' Ball, Maidenhead, 'Chrysalis', Apostle, Melon, Hexagon, Seal, Baluster, Diamond, Horsehoof, and Slip Top (the top of the stem cut off). You
can hope to acquire an example of this type, possibly Acorn, and
Hexagonal knop, maybe some others too, with patience—but
beware . . .

90

Fig. 28
Sixteenth-century spoons (Rack, early eighteenth). Gauntlet Seal; Lion Sejant; Maidenhead; Chrysalis Baluster; Apostle; Melon; Diamond; Latten Knopped Baluster; Hexagon; Baluster; Acorn; Simple Seal; Decorated Slip Top; Maidenhead. (All bear marks.)

The spoons at this time were bigger and more heavily made. The stem is heavier, and the hexagonal stem was hammered flatter, no doubt to give strength to the rather unsuitable medium. It is as well to study spoons carefully, for they are the one category of pewter that one can confidently expect to see of the sixteenth or even fifteenth century—but this will probably be in a specialised dealer's shop, or major saleroom, and museums. The big heavy-handled round-bowl spoons one occasionally sees are Dutch, easily distinguished, and of much later date. Hilton Price pioneered the subject of pewter spoons (and latten) in his *Old Base Metal Spoons*. In *The Connoisseur* issues of April and July 1970, 'Latten Spoons', I have traced the development of the spoon, and drawings help in dating at a glance, for the types are almost equally applicable to pewter and the two media are considered together. It was wryly amusing to read in *British Pewter* that one could 'make a collection of many hundreds'. One would indeed need a very long pocket, and a life to match.

Spoon-makers' marks are delightful, from beautifully made irons, being small beaded circles with initials and emblems. But trouble and competition suddenly sprang up, and were tackled by some pewterers joining the new faction in the second half of the century. Latten (brass) was now allowed to be made in this country, and it must have been realised that it was a far more satisfactory medium for the leverage imposed on spoon stems. Some pewter spoon-makers took to making latten spoons—and ructions followed (see *The Connoisseur* articles mentioned above). This appeared to take place at a critical time for the Company. There had been many reports of misdemeanours; the Company had difficulty in maintaining control (makers were always finding fresh ways to kick over the traces) and it was a bitter struggle—which in effect the Company lost, for latten spoons became the larger proportion of the examples which have survived from this period.

6. 1600–1710

Part I. 1600–60

Specimens prior to 1660 are still very rare, with the exception of a newcomer series—church flagons. The Church had been despoiled—particularly of the silver plate—and thus impoverished it was allowed in 1603 to use pewter for flagons, for bringing the wine to the table. So starts a really fine series of the most dignified ware in pewter—the rather rapidly changing styles of church flagons. Probably most styles are known by those interested in pewter, so let us seek certain points. The first type ('1605' type) ran till about 1625, and is very solidly made, of simple and most effective design, slightly tapering drum with a little entasis. The base spreads a little, and the lid is faintly like a skull cap, with knop surmounting. The thumbpiece is the most remarkable item, being an adaptation of a German design—solid, towering. The bottom of the container drum is curved, and does not touch the table (Plate 6). It was made in various sizes, and very seldom carried a mark. Those marks I have seen are on the back of the thumbpiece. This type is supplanted quite suddenly and universally, as though the Company ordained it, by the 'Bun Lid', in which the lid is waisted and has no knop. There is almost always a touch mark on the lower part of the handle. A knop was soon added and at the same time the bottom of the container was a

93

disc filling in the bottom of the base, i.e. flush with the table, on which the maker now put his mark. It has a rather weak-design straight-topped thumbpiece with a heart-shaped piercing (Plate 6). A great many specimens exist in churches and in collections. A vast quantity of domestic pewter was made at this time, most of which has perished, but church flagons have been preserved by inactivity and security in the Church. They may not be disposed of without a faculty, but few in collections can be matched by the document to prove that such permission was ever given. We are not sure if the same styles were used domestically, but quite frequently domestic tankards, both flat-lid and dome-lid, were taken into use by the Church as flagons. It is unfortunate that there seems to be so little opportunity for the Church to display such treasures with safety, and they command more interest in collections where they can be seen and used in research and study. Presumably each flagon had a plate on which it stood. These, when specially made, have a very short bouge, and are now very rare. They are about 7″ in diameter.

Plates proved to be of the early seventeenth century appear to be almost non-existent, but, as mentioned in the previous chapter, it is probable that many bumpy-bottom plates considered to be sixteenth century are, in fact, of the first quarter of this century, particularly those with slightly wider rims, for this was probably the style in fashion from 1600 to 1650, with the rim diminishing in size latterly. Dishes with rather deep wells followed, and have a fairly narrow rim—of about 1″—and a gently sloping

Fig. 29
Bowl-dish (see Fig. 9). Mark, cocks with 1646. c. 1650. Diameter 15⅞″.

bouge. One in my possession bears a touch dated 1646, and is probably pre-1660. Two plates which I take to be of this period have 1″ wide rims, and gentle bouge, with the bottom of the plate not so much bumpy as crowned. One is a gentle dome, the other a very flat cone. Both have 'hall marks' on the rim and are therefore about 1630–50. The deductions of dating are simple and obvious —the bumpy bottom tailed off when striking 'hall marks' started c. 1630: the gentle bouge and plain 1″ rim together usually occur

Fig. 30
Bumpy-bottom plates of seventeenth century. (a) With gently domed well. No touch mark, but 'hall marks' of 'A.I.' ('Cott. O.P.' 5700). c. 1645. Diameter 9¼″. (b) With low conical well. No touch mark, but 'hall marks'—all four—are of a head and shoulders. C. 1635. Diameter 8½″.

with late bumpy-bottom plates before very broad-rim plates and the more intricate reedings appear.

During this period porringers continued, and by now the sides were straight, with one ear for a handle. Of salts we know nothing; of candlesticks little, they are rare and seemingly trumpet-shaped at first; then, or perhaps concurrently, with heavily knopped stem, a drip tray, and rather low bell base. There is one class of decoration, finely cast, of which the Granger candlestick at the Victoria and Albert is a superb example, dated 1616. This decoration otherwise is preserved in a very few goblets and cups of apparently a very narrow period c. 1616.

The series of baluster measures is uninterrupted. In the six-teenth century they had borne a ball, and some a hammerhead as thumbpiece. Most collectors, myself foremost, do not accept that a 'wedge' type was made with no thumbpiece—maintaining that it is the strut leading to a missing thumbpiece. Surely it needs a firm leverhold ? How would stubby thumbs have coped ? Every single other type of lidded vessel is set up with a very imposing thumbpiece. No, a 'wedge' is an emasculated 'Ball', or a hammer-less Hammerhead. The earlier balusters were thin and slim, and are more likely to have been inspired by medieval pottery than by the leather bottle. So they followed the lines—slim, simple and gentle-curved, not flaring out at the base in very early specimens.

Fig. 31
Balusters of the early and late seventeenth century. (a) 'Wedge', quart, probably originally bore a ball thumbpiece, very gentle lines. Has a branded house mark in the base (see Fig. 14). Mark 'D.B.' with heart. c. 1630. Height to lip $7\frac{3}{4}''$. (b) Ham-merhead, gill. Much fuller lines. See rather weak hammer. Capacity check mark AR. (Anna Regina). Mark 'R.W.' c. 1700. Height to lip, $3\frac{1}{8}''$.

The bottom handle attachment fitted very flush with the body, and the mark was usually on the lip. But now they swelled, and grew a wider base giving more stability.

The measure was for wine or spirits, and was 'Old English Wine Measure', which is five-sixths Imperial (which is still the standard fluid measure in the U.S.A.). We have dealt with the house marks sometimes found on them and often faked. The Ball dropped out of use except in a very rare type, but the Hammer-head continued into Queen Anne's reign. In the meantime, in the middle of the century, the body became squatter and more curvaceous. It is not enough to judge the date, as one writer has said, by thumbpiece type—one Hammerhead was made about 1780, others presumed before 1550, but for these two types at least it is far better to judge by body type. Beware these early

96

fascinating basic types—there are more fake pre-1650 balusters than genuine.

The spoon-makers were in obvious trouble, to judge by the comparatively few specimens left to us. Whereas the church flagons particularly show us superb design, craftsmanship and finish, the spoons had become weak and crude in design and execution by comparison with those of the previous century. They were assailed by the far more practical latten spoons. The latten spoon-makers —some of them known to have been pewterers—now piled on a

Fig. 32
Very weakly formed knops of late provincial spoons of seventeenth century. (a) Very poor Baluster knop. c. 1650. This bears a cross on the bowl, said to protect from the evil eye. (b) Possibly a most unsuccessful attempt to cast a simplified Maidenhead. No mark (no wonder!). (c) No more than a Blob knop. (d) A very strong Slip Top. Note flattened stem. Mark 'R.A.' with two keys, 'Cott. *O.P.*' 5393—wrongly attributed to sixteenth century. c. 1640. Length 6⅞".

mortal blow: they tinned their yellowy-gold brass spoons, and turned out ostensibly pewter (or silver?) spoons of great hardness and resilience. You may come across a latten spoon still completely covered with tinning, but more usually showing only traces. The types of pewter spoon you may encounter are Baluster, Seal, possibly late Maidenheads, and of course pre-eminently Slip Top. These bear no knop, the stem appearing to have been cut off on the slant, like a pruning cut. Stems in this period are very much flatter and bowls very much more round. Odd examples of extremely debased but recognisable styles belong to this period.

Part II. 1660–1710

This is the narrow period when an enormous amount of the most attractive types of pewter was made, and has been preserved. All pewter of this period is irresistibly attractive and forms the greater part of established collections, and the ideal of newer collectors.

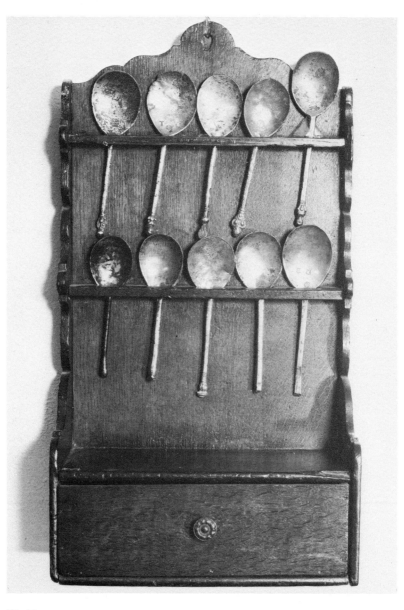

Fig. 33
Seventeenth-century spoons. Maidenhead;
Maidenhead; Horse-hoof; Alderman (or
Master of the Pewterers?); Maidenhead;
Blob; Weak Maidenhead: Weak Baluster;
Slip Top; Slip Top with two minute nicks
at end—the foetus of the trifid.

It would certainly seem that the flamboyant enterprise with which the makers threw off the heavy control of the Company, and the stimulating effect on design of a serious trade recession, unleashed an unbridled freedom which was exercised with the utmost good taste.

Emerging free enterprise was spurred on by keen competition, not only because of increased output of pewter, but also from the growing intrusions of pottery, and possibly silver, which was just within the reach of some of the pewter clientele.

So wide is the field, and it is so necessary to cover it adequately that it is difficult to know where to begin and what to omit. Certainly some inevitable omissions as to type, styles or details are bound to disappoint some readers.

Church flagons continued to change in style, even more dramatically. The first type after the Restoration is perhaps the best known of all—the 'Beefeater' (from the similarity of lid to cap). The base is widened considerably, giving a lower centre of gravity when filled, and a greater stability. The thumbpiece is usually a pair of hemispheres, or 'twin cusp', although in the western part of the country, and/or perhaps a little later, a form of twin kidney with projection was popular. The mark was usually inside the base, and for the first time we find 'hall marks' on the lid. This type is very pleasing and probably occurs most often, and was made by many different makers (Plate 7). A most handsome type which is seldom seen, most of the eight or nine I can call to mind being in private hands, was made by only one maker—John Emes, 1686. His touch is on the handle. The drums always show fine hammering, his handles are always weak, and he experimented with different types of thumbpieces (Plate 7).

At the end of the century flagons took on a new look, being of greater diameter and marked tapering. Belatedly they adopted the flat lid (already well established on tankards, as we shall see below). The early shallow lids of tankards appear to be used later than the taller lids. Some drums carry two broad hoops on a more cylindrical body, others are plain (Plate 8). There is unfortunately insufficent accurate evidence for close dating. Too often details of makers are not known, and where sometimes we have the date they struck their touches, seldom do we have their terminal dates (death, or ceasing manufacture).

Spoons underwent a revolution in design. Puritanism had kept styles plain, presumably confined to the slip top. In place of a knop someone hammered the top end of a slip top, to flatten it

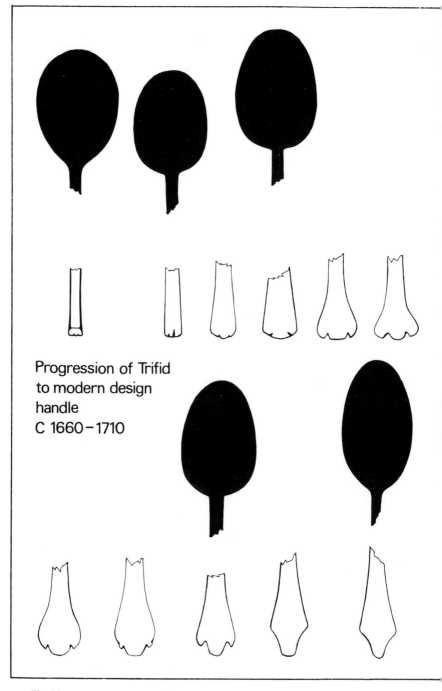

Progression of Trifid
to modern design
handle
C 1660–1710

Fig. 34
Rapid development of spoon handles, c. 1660–c. 1710.

and give a little leverage. Someone else went a step further and hammered it harder for greater flattening—and in doing so caused two splits at the top. Someone, whether in silver, pewter or latten, visualised the possibilities of developing design from this modest start. After the dull stark Puritan days everyone was ready to burst out with enterprising design. The spoons developed more oval bowls, flat wide stems, a finger running down the back of the bowl for support, and the top of the stem became a wide flat with a progression of detail—all emanating from the two nicks. It is particularly interesting that decoration and mantling grew around the edge of the top of the handle and the bare centre was used to bear

Fig. 35
Royal portrait spoon, Queen Anne, with one recently cast by and given to me by Mr R. F. Michaelis—both cast from the identical mould. Length 6⅞″.

successively William and Mary, and Anne. It is strange that although spoons in both latten and silver bear the surrounding decoration, in those two media it is left blank in the middle, and only in pewter does the royal portrait occur.

A moment's thought shows how logical it is to take porringers next. If you were well brought up you used your spoon in a porringer. It puzzles me why the porringers of the earlier part of the century had been straight-sided—and as a corollary, why spoons had become round-bowled. Perhaps the porringer makers realised that there was something amiss in their liaison with the spoon-makers, for from 1650 the curve in the wall was reintroduced. There may be a little confusion over the term 'porringer'. Its use was for eating any porridge-like substance, such as a thick soup, or stew. However, what we call a 'Two Handled Cup' or a 'Loving Cup', the silver lovers call a 'Porringer', and they refer to our porringer style as a 'Wine Taster' or a 'Bleeding Bowl'. I wonder how much wine tasting was undertaken, and how many receptacles for blood-letting were of silver, not graduated? Michaelis dealt with porringers in great detail in *Apollo*, 'English Pewter Porringers', July, August, September, October 1949. They developed in small details of casting, but almost all bear roughly triangular-cast fretted handles.

Fig. 36
Very pretty porringer, unblemished in shape and condition. Mark 'W.B.' c. 1695. Diameter of bowl $5\frac{1}{8}''$.

Various designs on the handle are their greatest attraction —but more than one or two specimens together do not display very well. An unaccountably rare class of domestic ware is salts. The pewterers drummed up business in many types of ware by rapid changes of style—probably none more so than salts. Therefore presumably comparatively few were made of each style. Rare indeed they are . . . but you never know!

Salts, like spoons, are well worth studying in other media— for instance silver, and pottery for salts. In the preceding centuries very great store was set in the ceremonial use of salt. A standing salt was placed at the host's right, and two smaller standing salts and two trenchers at the ends of the table. Trencher

102

salts (less than 1″ high) were laid down the lower tables. After grace, a napkin was removed from the principal salt, and salt was taken as a ceremony. The trenchers were for use, and the salt was taken out with a knife. The food was not dipped into the 'cellar' (corruption of French *salière*). After the meal everything was cleared away except the one principal salt. The silver specimens of the sixteenth century were magnificent complicated structures, and were very easily damaged. The Puritans suppressed display and ceremonial, and from the middle of the century the honoured

Fig. 37

Salts. (*a*) Base section of Bell salt. Although a salt in itself, this is the base of a lighthouse-shaped composite whole comprising two or three 'storeys'. Mark $\frac{H}{W}$. c. 1600. Height $3\frac{1}{8}″$. (*b*) Collar salt. No mark. c. 1660. Height $1\frac{3}{4}″$. (*c*) Plain capstan. No mark. c. 1680. Height $2\frac{5}{8}″$. (*d*) Gadrooned capstan. Mark 'T.L.' c. 1695. Height $2\frac{3}{4}″$. (*e*) Bulbous trencher. Mark illegible. c. 1700. Height $1\frac{3}{4}″$. (*f*) Ringed bulbous. Mark G. Lowes ('Cott. *O.P.*' 3001). c. 1720. Height $1\frac{1}{2}″$. (*g*) Trencher. No mark. c. 1720. Height $1\frac{1}{4}″$.

guests who had been 'above the salt' now dined in a separate room. Apart from the base of the Bell salt previously referred to we have no salts until the very attractive wide octagonal Collar salts of about 1660, of which very few genuine specimens still exist. They were soon outmoded by Capstans, which, plain, ran from about 1675, to gadrooned, up to about 1700. A completely new conception of design now appeared, smaller than Capstans, and instead of being waisted, were bulbous; the smaller, simpler pieces probably were the earlier. Note that the bowl of all these types is a separate well—the salt is not contained by the outside wall. Very broadly speaking, earlier salts have the smaller wells proportionate to the pieces.

Candlesticks follow salts if only because the moulds for casting the octagonal Collar salts were used sometimes for casting the base of the very much sought-after drip-tray type. These were made with a bewildering increase in complexity and additional means of ornament. They start with octagonal lip, drip-tray and base, the stem bearing bands of reeding or rings. Soon cast ornamentation of grapes and vines appears on the base. Then a welter of balusters and knops of exquisite lines are added. Beware—the price is high and fakes are frequent. Candle wax improved in quality towards the end of the century, and drip-trays fell out of use. Exactly where the knopped variety fits is not known for

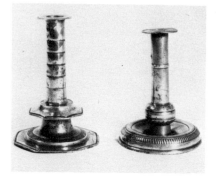

Fig. 38

Candlesticks. (a) The most attractive 'Drip-tray' design. The lines round the stem vary in spacing and number. Some, as this one, bear cast decoration of grape and vine on the base. No mark (if any it would be on the lip). c. 1680. Height 6½". (b) The rarer knopped 'stick'. Occasionally gadrooned, as example. No mark. c. 1695. Height 5¾".

certain. Perhaps they were of the same period. Certainly some have gadrooned edges to the base, placing them about 1700. Perhaps the knop followed the abandonment of the drip-tray. One illustration in two different books by the same author is confusingly dated 1675 and 1695 respectively. The latter must be the nearer. There is a further complication with this type—in some the stem stands up tapering out of the base; in others the stem grows out of a well in the base. These have not been compacted downwards—if that had been the case wrinkles and fractures of the metal would show. When marked, the touch is on the top of the lip on all types.

Plates and dishes are an obvious essential to any display, and this period sees the biggest range, changes, and prettiest styles of the whole history of pewter, some running concurrently. We left the early part of the century with bumpy-bottom plates and dishes, and rather bowl-like dishes. After the Restoration (or possibly a

Fig. 39
Broad-rim plate. More usually plain rimmed. The wells are very shallow. Some were used as patens, but are more likely to be of domestic use. Mark 'W.S.' and crown, and 'hall marks'. 'Cott. *O.P.*' 5961. c. 1645. Diameter 9⅝".

little earlier) the ultra broad-rim dishes and plates were made, *some* being used as patens, but mostly for domestic purposes. It is difficult to imagine these as everyday plates for meals, as the well is so small and shallow. Nevertheless, they show many knife cuts. What was the day-by-day table tableware from about 1660 to 1675 ? Perhaps they were the dishes with rather broad rims and either incised rings or deeply grooved rings—but plates of either

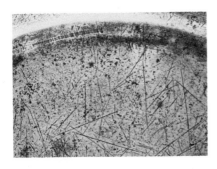

Fig. 40
Close-up of knife-marks. (Late seventeenth century.)

type are very rare. About 1675 the two types 'Triple Reed' and 'Narrow Rim' appear. These have from two to four rings or reeds *cast* near the edge. Triple-reed dishes are the most likely type of seventeenth-century ware you may encounter; narrow-rim dishes are exceptionally rare. It is very tempting indeed to make out a case to show that narrow-rim plates started with bold reeds, and gradually grew a plain band inside the reeding, on which 'hall marks' and owner's initials were placed. It grew until this style is almost indistinguishable from triple-reed plates. The difficulty is

105

Fig. 41

(a) Broad-rim charger. Maker S. Jackson ('Cott. *O.P.*' 5741, *L.T.P.* 11). c. 1655. Diameter 18¼", rim 3½". (b) Scalloped decoration on broad-rim charger. Carried out by three punches —a star, a fleur-de-lys, and a crescent. Mark 'R.S.' c. 1670. Diameter 20½". (See Fig. 44.) (Very similar decoration was in vogue for a very short time c. 1585. I do not know of another example of seventeenth century.) (c) Triple-grooved charger. The broad rim was at first decorated with deep grooves. Very rare. Maker G. Smith ('Cott. *O.P.*' 4347, *L.T.P.* 353). c. 1670 Diameter 18¼", rim 3". (d) Triple-reed dish. Less broad rim. Also by G. Smith. c. 1690. Diameter 16⅝".

Fig. 42
There seems to be a strong case for typological development of the narrow-rim
plate to the accepted triple-reed plate
(*extreme right*)—except that the evidence of
the average of the makers' dates tends to
show parallel existence. Tantalisingly,
there is no close dating on any narrow-
rim or triple-reed plate, except when
decorated as a Coronation souvenir. I do
not recall having seen a triple-reed plate
used for William and Mary's coronation
decoration.

that narrow-rim plates are usually provincially made and, from
what good evidence we have, tend to be slightly later on average
than triple-reed plates. 'Hall marks' were placed in the well of
the very narrow-rim plates, otherwise on the face of the rim. At
the turn of the century the rim was greatly simplified—with only
one ring—the 'Single Reed'—still with 'hall marks' on the front

Fig. 43
The single reed—in this case wriggled. The
single reed ran from c. 1700 to c. 1730.
(When acquired this plate had an extremely thick scale, chipped in different
layers, and the wriggling was only detect-
able by close inspection in good light.)
Mark and 'hall marks' P. Angel ('Cott.
O.P.' 94). c. 1700. Diameter 13". (See Fig.
45.) (The owner's initials $\begin{smallmatrix} H \\ G\,M \end{smallmatrix}$ could have
been for 'Hill, George and Mary'.)

107

Fig. 44
Punched decoration. Occasionally plates and dishes bear punching as the sole decoration, more usually to augment wriggling. This is a close-up of Fig. 41*b*. Mark 'P.S.' and 'hall marks' four leopards' faces. c. 1680.

in the early years of the 1700s. Touch marks on triple-reed dishes and earlier types are on the back of the rim: thereafter on dishes and plates from c. 1660 they are on the back of the well.

Before leaving plates and dishes there is a large subject which must be introduced—Decoration. We have discussed the reedings on plates, dishes, candlesticks and salts; the cast decoration in vogue in 1616; the knops on spoons, in candlesticks and on flagons. All these were an integral part of manufacture. We now need to look at art applied after the piece was made, although probably applied by the actual pewterer. We have touched on the punched decoration which appeared in about 1585—this scalloping reappeared, very rarely, in c. 1685. 'Wriggling' swept the board in the whole of the latter half of the century. It probably existed in the previous century, for I have a plate of undoubted

Fig. 45
Close-up of wriggling showing the different sizes of tool used. Designs were traced, and the lines followed by 'walking' a screwdriver-like tool, corner to corner. It is very easy to carry out. (See Fig. 43.)

Tudor (or earlier) period, which bears traces—but it does not seem to have caught on until the Restoration. Then suddenly there appear the big wonderfully decorated 'Carolus Rex' dishes (Victoria and Albert, and Wisbech Museums). There are many in existence, very prized, and possibly some are fakes. (I have never recognised one as such, but most have been much cleaned.) 'Wriggling' is, I suppose one could say, a form of engraving, but instead of cutting the lines deeply (and weakening the pewter) they are formed by 'waddling' a chisel along, pivoting on its corners at each step, zigzag. The design was traced, and different-sized tools were used, and varying pressures applied. As well as on Restoration thanksgiving and commemorative pieces, it was used for

Fig. 46
Coronation souvenir of William and Mary. The skill of the wrigglers vary considerably, and was probably done by the pewterer himself. Wriggling was used on plates, dishes, and flat-lid tankards chiefly. Mark 'R.S.' and crown ('Cott. *O.P.*' 5940, *L.T.P.* 165). 1692. Diameter 8½".

decoration, particularly on narrow-rim plates and flat-lid tankards, and showed the contemporary forms of flowers, birds and busts. Its maximum use was for coronation souvenirs of William and Mary. In this connection it is amusing to note that coronation souvenirs were as trashy then as at the last few coronations—the pewter in the plate illustrated is little better than dutch cheese. Wriggling carried on beyond the end of our present study period. Engraving was the normal treatment for arms, crests and inscriptions.

Wriggling on flat-lid tankards! No doubt a general description is unnecessary—probably everyone who knows what pewter is would like above all to have a decorated flat-lid. It would not deter them to know that when plain they are more rare. Perhaps the latter were so little regarded that they were traded in, while

Fig. 47
Flat-lid tankard. They are more usually decorated than plain. Ramshorn thumbpiece. Bears the AR capacity stamp. Note the concave curve of the drum—the opposite of entasis—to give a slender effect. Maker Henry Seegòod, of King's Lynn ('Cott. *O.P.*' 4169). c. 1695. Height 5¼″ to lip.

the wriggled one was kept carefully. There are fakes, and reprotions—so, as usual, take care. The early shallow lid c. 1655 with single tongue projection, insignificant base, over-curved handle, twin-cusp thumbpiece, and squat drum is virtually non-existent. The middle period (c. 1675–85) have a little depth in the lid, and a handle with extension dropping from the top to give a good attachment to the body. The thumbpiece is possibly a 'ram's horn', or twin crescents backing. There are probably five projections in front of the lip of the lid, and these are perforated. 'Hall marks' are on the lid, and touch marks in the base.

Subsequently, the base is heavier, the handle more solid, and butts onto the body. No doubt the fragility of the perforated projections (or denticulations) proved itself, for they now become solid. The body becomes either more squat or taller and thinner. Whereas earlier drums had a marked entasis (outward bowing), which conveys parallelism to the eye, they are now straight and tapering—sometimes even convex. Thumbpieces are either ram's horn, twin cusp or twin love birds. It is thought that the type ran

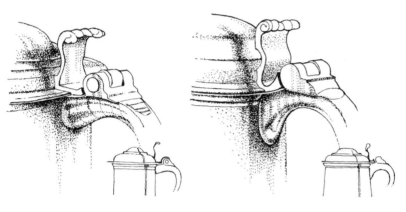

Fig. 48
The earlier (c. 1685–c. 1710) handle attachment of dome-lid tankards, butting onto the drum; followed by a more secure 'beak' running down the drum, c. 1700+.

until just after 1700. Latterly they may have been made without the denticulation—or, having been damaged, it was filed off contemporarily.

It is generally thought, erroneously, that dome-lid tankards followed after flat-lids. They were certainly made in 1685, for we have seen that Touch No. 420. L.T.P. ('Cott. *O.P.*' 5930) depicts a dome lid and '85'. The remarks as to body, handle and thumbpiece apply equally to dome-lids as to flat-lids, except that quite early dome-lids adopted an approximation to a clenched fist as thumbpiece. Dome-lids are best dated by body and handle type. In the period under review, at first they had a high dome and small squat body, with small base—and *may* be earlier than 1685; then a

111

Fig. 49
The scroll thumbpiece, which was first
used c. 1690, and continued to c. 1750.

little lower dome, over-wide brim to the lid and body with marked entasis. The handle on both of these types runs down the body. The denticulations are bolder than on the flat lid. Both bodies are plain—free of encircling rings (Plate 9). In some cases these purely joyous purpose items have been taken into use in small parishes as flagons.

It is puzzling that almost all containers of liquid appear to have been automatically provided with a lid. One needs to go back to Romano-British times to find lidless containers—the ewers and flagons. *All* subsequent examples—flagons, tankards and measures —have lids. Hygiene and safety of contents were probably not the reason. Perhaps the Company ordained that lids were to be made, to keep the price up. Perhaps it was further evidence of the rebellious enterprise of the 1680s, in the face of a trade slump, which led to the production of the lidless 'tavern pots'. Their design differs from the lidded tankards, being tall, slightly concave sides, with two bold bands or hoops round the body, and usually bearing an engraved inscription showing ownership (Plate 9). Those prior to 1700 are very scarce indeed, and those of the early years of the eighteenth century only slightly less scarce. These latter, smaller in capacity, have only one band, fairly heavy, or a fillet high on the drum, or two equally spaced light fillets, rather heavy handles, and appear to be engraved only very seldom. There is an even smaller group with gadrooning—grooves at an angle—

Plate 9 (a) Dome lid tankard. Note lid projections, and entasis. Mark, "W.W." (Cott. O.P. 6028) Ht. to lip, 5″. (b) Lidless tankard, or "tavern pot". Mark, "I.D." (Cott. O.P. 422) c.1685. Ht, 6⅜″. It bears engraved ownership.

Plate 10   Very dignified Spire flagon. Varieties ranged from c.1715–c.1770. Note 'broken' handle. No maker's mark – "Hall mark" "H.I." c.1760. Height 12¼″ to top.

round the lower part of the drum. A few years ago one of these was found hanging up with nineteenth-century tankards—and at the same price.

The last category to be dealt with is measures—baluster measures—perhaps the most important and interesting group to understand fully. They are pre-eminently suitable in design and function for pewter, and do not appear in other media—certainly not silver, although I have seen one of later date in brass. We have

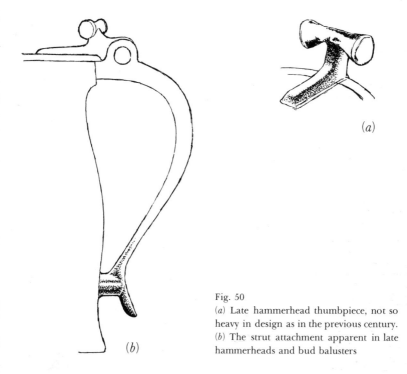

Fig. 50
(a) Late hammerhead thumbpiece, not so heavy in design as in the previous century.
(b) The strut attachment apparent in late hammerheads and bud balusters

already met them in the earlier periods—as Ball, and Hammer-head. The latter continued to be made after 1660, apparently up to the end of our period, since one I have bears AR and was probably stamped at the maker's in her reign.

By now the hammer was usually very thin and emaciated, and whereas formerly the lower handle attachment was flush to the body, now it stands off, connected by a strut. Formerly tall, slim and of little curvature, by 1660 they are invariably more squat with fuller curves, including the base flaring out. The original three sizes recorded in 1480 as 'Pottell' (half gallon), 'Quarte'

113

and 'Peingte' had been joined by 'Halfe peingte' in 1556 records, and now by gill and half gill. (Be well warned to avoid a quarter gill, however twee and tempting.)

The series of balusters must have run for 350 years or more—starting at latest c. 1480 and ending in c. 1820. By about 1670 the 'Bud' thumbpiece makes its entry, so called for want of a more accurate descriptive name. Each end is a 'leafed' projection faintly like a bud, and comparable with car springs dressed together. These are most desirable and not unduly rare, and specialised dealers may be able to show one now and then. It is almost true to say that no two buds are alike, for even when two come from the same mould they will have different grooves cut round the body. Like humans, some are tall, some are solid, and they carry their curves in different ways. Grooving bands were very often used, but only on balusters with this style of thumbpiece.

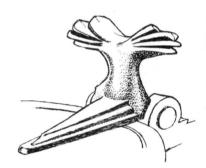

Fig. 51
Bud thumbpiece, also slightly reminiscent
of car spring-leaves. c. 1670–c. 1780.

Some have plain bodies—but the nuances of different curves of the outline are limitless. The strut attachment of the handle is always evident and handles are heavier than in the slim earlier measures. The lids, too, bear rings of varying number and distance apart. The thumbpiece sometimes bears a little pip in front, which also occurs on some later hammerheads (and so can be taken as evidence of the late seventeenth century). The thumbpiece is attached to the lid by a three-tiered flat wedge. The touch marks in early specimens are small and may be either on lip or lid, or, particularly on late examples, there may be no mark. At first slim and plain, curves and rings developed. Occasionally they were made without lids. It is surprising to find two touches on the London Touch Plate—Nos. 362 and 584 both of which depict a lidless baluster of bud form. These would have been struck c. 1682 and c. 1700 respectively.

114

Fig. 52
There is more variety in line and detail in bud balusters than probably any other measure, as this small selection shows.

With so much styling variation one would think them easy to date. Unfortunately, it is the very opposite. Few bear dated touches, which would give only earliest possible dates. Few are by recorded makers—and dates of deaths are scarce. With experience one can feel fairly confident of dating some to within ten years. Sometimes luck may help—if a bud bears a touch with date say 1680, and bears AR crowned on the lid, the piece is not likely to be more than forty years later than striking. Therefore the latest date would be c. 1720. But AR would not have been used before 1702—so its dating is 1702–20. If you have a bud with a pip on the thumbpiece, and/or the handle very closely attached to the body, then it will be early—c. 1680–1700. It would be more misleading than helpful to try to formulate any rules of dating by design. Of all types, these appear to be subject to the fewest rules in details of style (and capacity). For this reason it will be more convenient to take this class beyond our terminal date of 1710. The capacities had been laid down, but the variations within a given size are enormous—up to about ± 10 per cent.

Dr Homer and I have at different times carried out wide surveys of buds to try to find reasons for the erratic departures from the standard, but we have had no real success. Provenance is invariably unknown, date difficult to define, and the result means nothing. It was not intentional short measure, for most are over-standard. One interesting clue was three (quart, pint and half pint) by different makers, and of different styles; each bore the same owner's initial's. They are exactly in proportion.

It should be pointed out here, as well as later, that some 'short measure' buds conform exactly to the Scottish measure, and the possibility that these may be the earliest known Scottish measures has escaped previous writers.

Buds appear to be discontinued by about 1760. With great reluctance and nostalgia we leave buds, and the seventeenth century.

7. 1710–1820

We left trade in a bad state, and have seen the reasons for the spate of new ideas which were rife from 1660 to 1720. Anyone aged eighty in 1720 must have been subjected to changes of designs far more extreme and sudden than we complain of today. Trade continued to be bad; there were many insolvencies. The Company was tottering, and its history of weakness and vacillation came home to roost. It was disregarded. With the coming of the House of Hanover, designs became Teutonically heavy, and largely gross and mundane. Inspiration and attempt to float new styles deserted the pewterers. Instead of meeting competition by upgrading design, they kept costs down by using the same plant for a long time, and their moulds had to pay for themselves over and over. Decline was general. There were some extremely successful pewterers— William Eddon and Thos. Chamberlain, for instance—but only a comparatively small proportion. Pewter faced competition from pottery, porcelain, silver and, later, plated goods and other base-metal alloys.

Whereas up to about 1750 items in collections could almost be checked off against an inventory, with only a few blank lines for the very unusual 'unknowns' which have not appeared pictorially in books, there is now a wide range of miscellanea—you may meet a number of awkward trifles of very indeterminate date. It is really

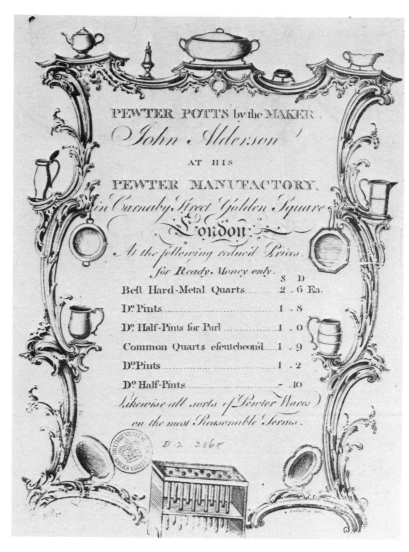

PEWTER POTTS by the MAKER
John Alderson
AT HIS
PEWTER MANUFACTORY,
in Carnaby Street Golden Square
London
At the following reduced Prices
for Ready Money only.

	S	D
Best Hard-Metal Quarts	2	6 Ea.
Dᵒ Pints	1	8
Dᵒ Half-Pints for Purl	1	0
Common Quarts escutcheon'd	1	9
Dᵒ Pints	1	2
Dᵒ Half-Pints	-	10

Likewise all sorts of Pewter Wares
on the most Reasonable Terms.

D 2 2868

Fig. 53
Pewterers' trade cards. (*a*) John Alderson,
who struck his touch in 1764. (*b*) William
Life, whose date attributed by Cotterell
as c. 1700 is obviously an error. c. 1780–
c. 1800. (The ranges depicted are remarkably varied, and illustrate types generally
assumed to be much later—or not made
in pewter at all.)

(*By arrangement with the British Museum*)

118

Wᵐ LIFE,
Nᵒ 87. Golden=Lane Old Street.
Sᵗ LUKES,
SELLS all sorts of Pewter Ware on the Most
reasonable Terms Quarts from 16 to 19 Per Dozᵉ
Pints from 10 to 14 Per Dozen Best.
BAR Quarts 2..6 to 3 Each
Best BAR Pints 1..6 to 2 Dᵒ
Orders per penny Post duly
EXECUTED

surprising how much earlier many styles started than is usually thought, but wrong to ascribe the earliest possible date to pieces without evidence. They are more likely to have survived from their bulge of production, and later. A most useful reminder of this is to be seen in a few pewterers' trade cards (advertising hand-outs), to be seen in 'Cott. *O.P.*' Taken alphabetically (date of striking touch is not necessarily related to date of printing the trade card):

John Alderson (1764–82+), shown here, displays a teapot, sugar caster, tureen with lid, three-legged salt, sauce boat, spouted measure, barrel tankard with 'broken' handle, wavy-edge plate, candle mould, plate, tulip tankard, hot-water plate, very slender baluster. (Best hard-metal quarts were 2/6d. each, common quarts were 1/9d. each.)

R. P. Hodges (1772–1800+) shows a syringe, two cylindrical ink wells, a standish, toy watch, a very neo-classical two-handled lidded cup, and a doll's tea set.

John Kenrick (1737–54+) shows a plain-rim oval dish, a wavy-edge plate, and a heavy lidded tureen.

Richard King (1745–98) shows a dinner service of octagonal flatware.

William Life (Cotterell gives c. 1700 which is obviously an error—should be c. 1780), shown here, a caster, tulip tankard, footed sauce boats, oval teapot, beaker, cup salt, tankard with flattened handle attachment, and cream jug. He accepted 'orders, by 1d post, duly executed'.

Robert Piercey (1735–60) shows a picture of a neat pewter shop, visible in which are tea cups, porringer, candlestick, etc.

One would be very much inclined to place most of the items shown in these six trade cards in the nineteenth century, while the middle date of the working span of these makers is about 1767. So look twice at that Victorian tankard—occasionally one will be very much older. It is important, too, that we would not have expected many of these items to be made in pewter—e.g. three-legged salt, the two-handled lidded cup, footed sauce boats, and cream jug. Britannia metal we would expect—but here they are, advertised as pewterers' wares.

I have intentionally taken you for a ramble into unexpected products and dating, so that we can return to collecting in this period with an open mind, but sharper eyes—eyes to spot the early piece of a type which appears to be much later. By contrast, here are some items which may be met, which were presumably too commonplace to advertise on trade cards: church flagons, bowls, plates, footed plates, communion cups, porringers, bleeding bowls, tankards with lids (of various types), early-century lidless tankards, candlesticks, spoons, tea caddies, snuff boxes, chamber pots, tobacco stoppers, seals, whistles, medallions, mantel ornaments, to mention only some. There is plenty to go for; but the ability to recognise them is required.

Let us start with church pieces. We know that styles now seldom change. There are really only two basic types of English church flagon for the whole period. Initially a nearly-flat lidded type was produced, apparently in England first, but it was adopted almost universally in Scotland for probably 100 years, and by far the greater number of these were Scottish. This type soon had a successful competitor—the rather heavy first edition of the 'spire' flagons. The early type has a plain handle, and lowish dome

Fig. 54.
English flagons more usually Scottish. They fit into no developing sequence of design—quite odd men out. Used in England c. 1710–c. 1730, and in Scotland c. 1720–c. 1830. Mark and 'hall marks' illegible. c. 1720. Height 7¾″ to lip.

lid. However, it was soon made more slender, with higher dome, with knop, wider base, and, most important, a very well designed 'broken' handle (Plate 10). The total result is that this delicate, neat, always exceptionally well-made and beautifully proportioned piece stands out like a fairy amongst Hanoverian soldiers. To my mind it is one of the short-listed most attractive styles ever made—perhaps the winner.

Fig. 55
'York' flagons. (a) Straight-sided. More probably Lancastrian, and is the earlier. The lid, lid projections, and finial are c. 1715. Mark 'I.W.' (as 'Cott. *O.P.*' 6002—his drawing may be inaccurate). 9⅜″ to lip
(b) Acorn type. Magnificent example, engraved 'Bought . . . 1750'. Maker John Harrison, of York ('Cott. *O.P.*' 2162). Height 9½″ to lip.

Regionally there are two outstanding types of flagon—the 'Straight-sided York' and the 'Acorn York'. It may fan smoulderings of the Wars of the Roses, but I suggest that the straight-sided is more truly Lancastrian. These two seldom change hands, but are to be seen in collections. One should remember the fairly frequent use of domestic dome-lid tankards as church flagons. Many

121

exist in churches with inscriptions—'Ex dono . . .', etc. Other tankards and balusters marked with only the name of the village are perhaps village measures, not church flagons.

From flagons we pass to communion cups. They are very rare in the previous century (when they had small cups) and were omitted from this text. In our present period they are of large capacity, usually with the cup slightly tapering down towards the pillar of the stem. Most of those engraved are Nonconformist. The English specimens are much less frequent than Scottish.

To patens is the next step. Early in the century they were often narrow-rim plates mounted on a circular foot. Sometimes this was beaded, more often plain. After c. 1730 a normal plate stood duty for a paten, and several are inscribed as proof. Plates and bowls were used for alms and collecting. Some churches stood a bowl in each porch. One church in Norwich has four such narrow-edged flat-bottom bowls, engraved on the rim 'St. Andrews Church, North Porch', one for each of the other three porches. (There were a few reeded bowls in the seventeenth century, sometimes engraved, but very few outside the churches.)

Fig. 56

Plain-rim plate, bearing the house mark of the Corporation of Great Yarmouth. This type of rim is by far the most common to be met, and was in use from c. 1710 to c. 1820. No mark visible. c. 1760. Diameter 9".

Plates are almost always plain-rim of about 1" (the single-reed rim ran out by c. 1730) and there are also soup plates with similar rims of the late eighteenth century. The only difference at first sight between these and some of c.1660 is in the bouge. The earlier type is much more gentle in curve, and, correspondingly, the diameter of the bottom of the well of the later is much greater. For a short period c. 1760 there was a vogue for copying the French style of cast wavy reeded-edge plates and oval dishes. These are most attractive and break up a display of plain- or single-rim plates to great advantage. As we saw in the trade cards, octagonal plates were made, but are now very seldom in circulation. Another

Fig. 57
Wavy-edge dish. For a short time the
popular French style of wavy-edged plates
and dishes was adopted c. 1760. Maker
Thomas Chamberlain ('Cott. *O.P.*' 873).
Long diameter 18¾″.

Fig. 58
Lobed decoration. The famous Nuremberg brass or latten dishes were one of the Germanic
influences adopted under the Hanovers. These are quite usual in Corporation plate in
gold and silver, but are extremely rare in pewter. Maker Edward Leapidge ('Cott. *O.P.*'
2894). c. 1728.

rare class of dish of c. 1728—12″ to 16″—is of lobed decoration hammered out. Very few are known, but probably there are others in private hands, which may turn up in an out-of-the-way shop or a sale. They were probably made for decoration and ceremonial use, being like silver and gold rosewater dishes.

Identical in form with the church collecting bowls were those for domestic purposes. Obviously there were countless kitchen and household uses, but I think it interesting that a fine mahogany dressing table I possess has one drawer with circular cut-out containing a pewter bowl, exactly like the church bowls. This is by Aquila Dackombe, and reference to 'Cott. *O.P.*' shows that he was insolvent in 1761, and struck out of the Company in 1773.

Fig. 59
Typical bowl of the latter eighteenth century, with many uses, showing the actual piece of furniture in which it was found. Maker Aquila Dackombe ('Cott. *O.P.*' 1275). Diameter $9\frac{1}{8}$″.

Plates, dishes, bowls, to porringers. Like nearly all other products, porringers were very similar throughout the period—one fretted ear, on which the perforations remain unchanged. Many like to think them bleeding bowls, but the latter are always graduated in 4, 8, 12, 16 (oz.) rings. Oddly enough, these are the only British measures of any type giving 'read-off' markings. Whereas porringers are round-walled, bleeding bowls are straight-sided. Never willing to yield to the cant of antique dealings, who seem to think bleeding bowls very desirable, I do not see why this type should not have served equally well and far more often as kitchen measures. I like a piece, regardless of its possible use. These 'bleeding bowls' do not bear marks, and I take them to be late eighteenth century, with probably a long run.

124

Fig. 60
Bleeding bowl of the latter eighteenth century, showing 4, 8, 12, 16 oz. rings. Could it not equally be a kitchen measure? No other measures bear graduations. No mark. c. 1760. Diameter at top $5\frac{1}{4}''$.

Salts, candlesticks and spoons followed very different courses. Salts had reached the stage of low-sitting bulbous trenchers. After 1710 these became rectangular, with the corners cut off, in the well-known silver style. About 1750 a radical change occurred —they were made like a grapefruit cup, always without a mark. The container was the cup—single-skinned for the first time. These continued into the nineteenth century, but to our surprise we saw a three-legged salt prior to 1780 in the trade-card illustration. A

Fig. 61
Salts. (a) Rectangular trencher salt. No mark. c. 1720. (b) Cup salt. This type ran from c. 1750 to c. 1800. (c, d, e, f,) Developing styles of salts, probably in their typological sequence. No marks. c. 1780– c. 1820. Diameter of (b) 4".

125

great variety of minor differentiating outlines of small cup and trencher salts occurred around 1800—but in the absence of marks and other evidence we cannot get nearer in dating. It does not seem good enough to dismiss a range of household articles as 'c. 1800'—when one cannot be certain whether it could be 1780 or 1840—but that is the situation.

Candlesticks had a broken career. We left them rare and expensive in 1710–15. Excepting a very rare Newcastle type they disappear until about 1760! (There are some Dutch specimens of the period.) One can only presume that pottery and brass ousted pewter candlesticks. But they came back in full force with several

Fig. 62
Candlesticks. There were several rather similar styles c. 1770–c. 1820. No marks (ever). Height of (a) 8¾″.

designs involving knops, tear drops and balusters. The over-riding recognition mark is the iron push-rod for ejecting the candlestub. No marks are struck on these candlesticks.

Spoons—at last it seems that although they had been good enough for parents and grandparents, it was now realised that pewter spoons were not much good to anyone—except apparently the Dutch. (Dutch spoons are very heavy and gross, and the bowls are almost hemispherical.) What English spoons there are follow the slim styles well known in silver—the forerunners of Georgian and Victorian shape.

A very interesting historical link survives with Society flagons. In the middle of the century the workers in a trade used to get

Fig. 63

Society flagon. One of five or six different Societies in Norwich. Very heavy and late in appearance. Engraved with officials' names, and 'This Society was founded in 1819 at the Raven, King Street' (Raven Yard, King Street, Norwich, still stands, with a sixteenth-century dwelling alongside). Bears WR check stamp—contents 1 gallon. No mark. Height to lip 11½″.

together locally and pay into a fund for mutual insurance. In towns each major industry had its society. (From these development was natural, but in two different directions—Friendly Societies and Trade Unions). Each Society had one or more engraved flagons. Perhaps Norwich has more than other towns and cities—there are at least five different guilds' flagons existing. In that city the very strength and solidity of the Friendly Societies checked the development of the trade unions. In Norwich there were also the Mancroft Ringers—the bellringers at the famous church. Not only the flagon, but tankards too exist similarly engraved. Often these flagons are not engraved with the Society's function, but bear arms, and/or the supervisors' and headsmen's names (Plate 11). Their occupation (and thus the Society's function) can be found in Electoral Rolls.

Two-handled cups on a foot can be taken, broadly speaking, as of the latter half of the eighteenth century; those whose bowls sit at table level as the early half.

127

Leaving tankards to the last, we will deal with measures—baluster measures. Since we have already investigated buds for their whole run (due to difficulty in dating), one might think that only the double-volute thumbpiece type is left. That is by no means so. For instance, there are the bud-type bodies which were made lidless. The earlier are taller in proportion, the later (mid-eighteenth century) are very wide-rimmed, and the end of the handle is fitted flush to the body, like a soldier's hand as he stands to attention. The bud is the only type which was made lidless. It was followed by the double volute (so called as resembling the decoration on Ionic columns) in sizes from the gallon to half gill.

Fig. 64

Double volute balusters. (a) Pint O.E.W.S., with the descendant of a house mark engraved on the body—'The Sun'. (Although it is in Wine Standard, it is stamped with a portcullis and '00', perhaps short for 1800. Mark 'R.M.' ('Cott. O.P.' 5794). c. 1790 Height to lip 4⅞". (b) Half pint, with true house marks, also showing lid attachment. (On the quarter gill the fleur-de-lys is cast on solid sheet pewter.) Mark illegible. c. 1760. Height 4".

In this type the thumbpiece is attached to the lid by a fleur-de-lys (in the smallest size this is cast in sheet pewter). The handle, with ball terminal, is attached to a diamond-shaped plate on the plain body, for greater adhesion. Just occasionally a bud may bear one or two of these features, which is a strong pointer to dating. Conversely, early double volutes sometimes bear rings, and sometimes have a touch mark on the lip, and very occasionally house marks—one presumes their life ended in 1824 when the Act was passed establishing Imperial Measures, and abolishing Old English Wine Measure. I have never heard of one in Imperial capacity (remember they were measures for wines and spirits), although in Suffolk the lid was stripped off and an extra piece added to the lip to bring the capacity up. I have seen only the half pint, and the pint modified.

Double volutes appear never to bear W. R. crowned. It was used in other pieces in the first half of the century and right through the

Plate 11 Society tankard, engraved with the arms of the Wool Guild of Norwich, dated 1747. Maker, William Eddon (Cott. O.P. 1503. L.T.P. 470). Ht. $5\frac{1}{2}''$ to lip. Eddon must have been 73 when this was made. It bears W.R. crowned.

Plate 12 English Ball measures. (a) Half Gill, c.1680. (b) Half Gill. Mark "E.Q.", c.1720. (c) Half Gill, maker Hincham (Cott. O.P. 2329) c.1720. (d) Gill, ball very low, no strut. Mark "I.M." and 1662. Ht. 3⅜" to lip.

four Georges up to 1826. I have it on a Society flagon of 1819—eleven years before William IV. A very rare type is similar to a bud in body and conception, but the thumbpiece is a spray, but these I think may be Scottish (or Border country) and will be considered in that chapter—their capacity is between Scottish and Old English Wine!

Another very rare type, which I think may be North Country, but more probably is Scottish, is slender with ball thumbpiece. This, too, will be explained under Scottish measures. A further type, which no other writer appears to have commented upon, is

Fig. 65
Ale jug. Quart. A singular design in English pewter. Loosely, late eighteenth century. Seldom bear a maker's mark. Height to lip 6″.

undoubtedly English, and has a ball thumbpiece. This is attached at first by a ridged bar, then a flat wedge or bar. The handle, lid and body are typically bud (Plate 12). The first three specimens I found were all quarter gill, and I had assumed that they were used as the smallest size for buds, for genuine examples in this size of bud are extremely rare. One of mine is made by A. Hincham (c. 1710–50) and stamped W.R. crowned. However, I have since then come across a gill with touch dated 1662. Not truly measures, but containers, the pleasant rural-looking ale jugs are usually dated c. 1780–1810, and are found in four sizes, half gallon to half pint.

Under Measures I have omitted tankards which bear capacity check stamps, for as in the seventeenth century this guarantees only the contents, just as now some of our glass tankards are

129

Fig. 66
Measure, brass rimmed for protection,
pint. Pewter-rimmed half pint. Probably
c.c. 1800–c. 1830. Height of pint, $4\frac{3}{8}''$.

engraved with the capacity. However, one must in fairness dissoci-
ate one type of ostensible tankard which is more truly a measure—
that with either a brass or extra-heavy pewter lip. A glance at the
illustration will show great similarities to nineteenth-century
tankards—yet there are telltale differences—the fillet and base
reeding is different, and the handle finial, being a ball, is as on
double volutes, and was in fact used only throughout the
eighteenth century. Remember, the old Ale Measure was acceptable within the tolerance of Imperial Measure, so size alone is no
true evidence of date, as far as ale measure is concerned.

At first sight the subject of eighteenth-century tankards looks
very simple—just dome lids. A little further knowledge plunges
one into a jungle of body types, fillet positions, thumbpieces,
handle types and finials, further confused by many types appearing much earlier than generally accepted.

Let us tiptoe through some general facts to set the scene for a
simple overall analysis. Lidless tankards continued to be rare for
many decades. It is not until c. 1770 that lidless and lidded exchanged popularity. Almost immediately the proportion of lidded
became very small indeed. I wonder why: probably the soaring
prices of metal. There were more pewterers at work in the early

130

years of the century than at any other time (although lack of trade soon reduced the number). Yet to judge by existing examples lidless tankards were very scarce then—far scarcer than measures. Strange—but there are probably two very different reasons. Beer had become very expensive by the end of the seventeenth century, at 3d. a quart, and gin was cheap, strong, and popular: so much so that it is recorded that by the mid-century there were 7,000 gin shops in London (which advertised '. . . drunk for 1d., dead drunk for 2d.!'). Gin was much more popular than beer. Hence the many balusters, but few lidless tankards. In 1763 heavy duties and taxes were imposed on gin. The other reason is that probably in the last fifty years a large proportion of eighteenth-century pewter has been discarded due to its similarity to nineteenth century. It was not, and still is not, recognised. Most nineteenth-century styles started in the eighteenth century, and ran on for a long time. That is of course true of styles common to both centuries. Specimens of the eighteenth century are now far outnumbered by nineteenth-century examples—probably because many more were made after

Fig. 67
Lidless tulip tankard, pint ('tulip' for want of better description). (a) The style commenced c. 1720 with one fillet round the body (and has continued to modern times). Mark illegible—'hall marks' lion rampant four times. c. 1730. Height 4½".
(b) Quart. Maker Sir George Alderson ('Cott. O.P.' 38). c. 1820. Height 6¼".

Fig. 68
Range of tankards (mostly quarts), and various handle types, *all* of which could be either eighteenth or nineteenth century. Tallest $6\frac{1}{4}''$. Even this range is not comprehensive.

1825, with the advent of Imperial Measure, and cheaper beer, and because of the economic pressure on scrap metal prices in the 1780–1810 period.

This chapter relies largely on its pictorial presentation, with text giving the background. Let it suffice for description to say that the dome lid held a long reign of over 100 years (1680–1780+): that until about 1750 the squat tapering drum held the field almost on its own, with a sprinkling of 'tulip'-shaped tankards. Several other types appeared after c. 1760—all of which look remarkably like those of 100 years later. Moulds were kept in use for a long time, styles hardly varied, and dating is difficult. However, ancillary parts may help to narrow the field. It must be understood that at least four body styles were used at the same time. All the types appear to have had countrywide distribution, although in one of his books Michaelis describes the concave-sided type as confined to 'North Country and Scotland' and in another as 'Norfolk and Suffolk'. Perhaps initials in his notes (N & S) were misconstrued. In any case it appears widespread—certainly one is verified 'Devon'. The chapter on 'Marks' would enable anyone to record origins and, after seeing sufficient samples, to draw conclusions.

To assist easy reading of the graphic dating chart, some notes will be helpful. It must be emphasized that dating is still tentative. However, most dating can be substantiated by known date of death of a maker, of a feature, or associated evidence. It will of course be accepted that many minor details are smoothed over into general classifications. By noting dating of parts, one may be able to work down to a narrow dating margin, e.g. a chair-back thumbpiece in conjunction with a fishtail handle finial would probably be very near 1740. A drum with a fillet high on its drum with a 'Soldier at Attention' handle attachment . . . ? It would almost certainly be a made-up piece. To publish anything so undocumented as this chart is to take the risk that contrary evidence will soon be forthcoming—indeed, let us hope that it will be a stimulus to research, and that any proved errors will not have been seriously misleading.

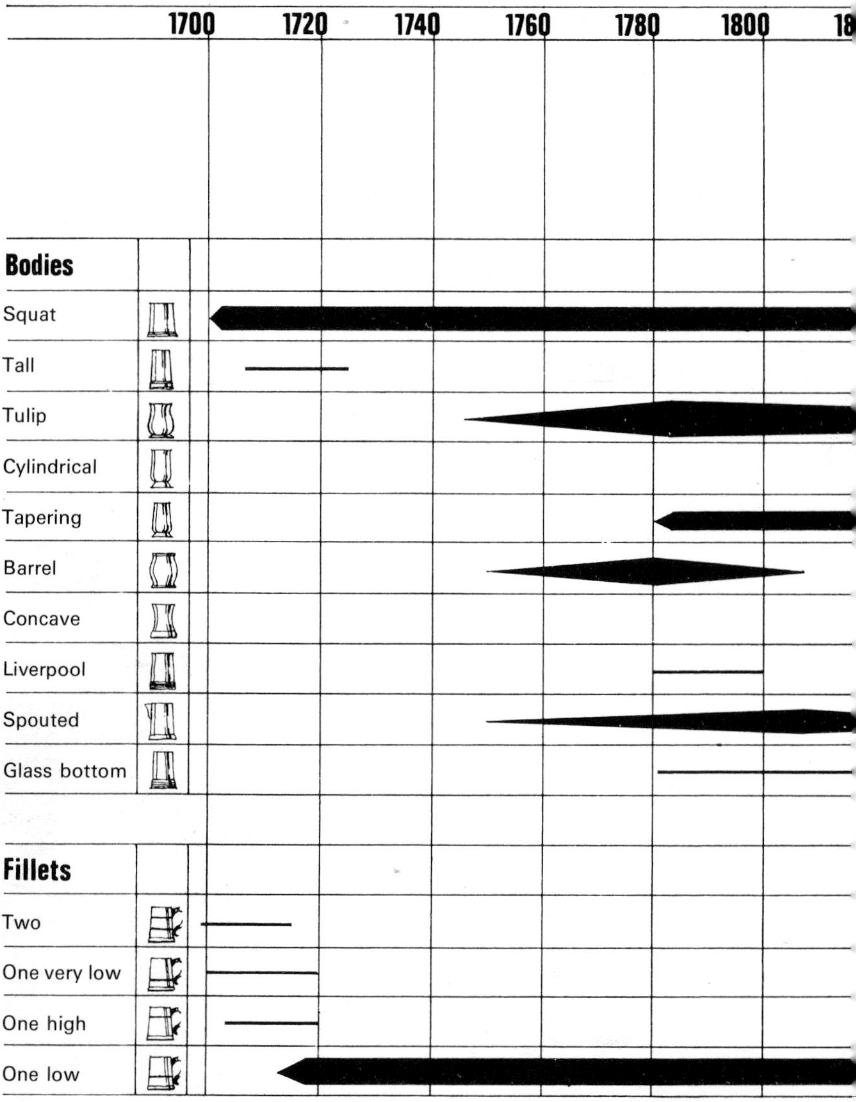

		1700	1720	1740	1760	1780	1800	18
Bodies								
Squat								
Tall								
Tulip								
Cylindrical								
Tapering								
Barrel								
Concave								
Liverpool								
Spouted								
Glass bottom								
Fillets								
Two								
One very low								
One high								
One low								

Fig. 69
Although dating is only approximate, by extracting the overlaps of dating of features one may be able to achieve fairly close dating. The width of band is a guide to quantity in existence.

		1700	1720	1740	1760	1780	1800	1820

Thumbpieces

		1700	1720	1740	1760	1780	1800	1820
Ramshorn								
Scroll								
Chair								
Pierced								
Open								

Handles

		1700	1720	1740	1760	1780	1800	1820
Solid								
Hollow								
Standard								
Broken								
Rectangular								
U and oval								
Attachments								

Finials

		1700	1720	1740	1760	1780	1800	1820
Heel								
Spade								
Fishtail								
Ball								
Attention								

135

8. 1820 Onwards

Porcelain, pottery, Britannia metal, brass, plated goods had successfully beaten pewter to its knees: but the alloy was still preeminent in the tavern bar. The pub was its home and saviour until a brief resurrection of decorative ware around the end of the century gave it a temporary boost. Small spirit measures were made in improved and hardened alloy and at this time more was made for decorative and presentation purposes. Although some other uses hung on, and a few more were exploited, the nineteenth century is the great rundown period from the eighteenth century, including its very moulds. It was held together only by the needs of the pubs—measures and tankards.

This is the period least researched in the whole history of pewter because of the disinterest born of plenty, for plenty there was until a very few years ago. Our exporters have done well, both for themselves and as dollar earners. But the pieces needed for comprehensive and easy research have sailed the seas.

The eighteenth century has, rather unexpectedly, stolen the little thunder remaining in the nineteenth century. The standby of the centuries—the balusters—ceased production c. 1820. Their successor we will deal with a little later, and was almost the sole 'new' style after 1820. All the tankards and their ancillary parts had been designed and produced by 1780. Few modifications developed.

Dishes, and plates particularly, survive in large numbers up to about 1820, culminating in the gigantic coronation service of George IV in 1821. Pieces of all tableware, including tureens, were pillaged after the feast in a disgraceful exhibition of mob greed, and are now widely distributed. Another widely spread service of a little earlier bears the Great Yarmouth coat of arms (three herrings) stamped like a house mark, which in fact it is. (I once had a quart, pint and half-pint double volute bearing a smaller Yarmouth stamp.) Many classes of ware turned over to Britannia metal, such as all church pieces. Ale jugs may have continued to be made, but in the absence of proof one would think not.

Undoubtedly the extra heavy-lipped tankard/measures continued to be made, but the type which has persisted in one size even until this present decade is the bellied measures. Much stouter in its major diameter than the balusters, it is otherwise of the same conception—but it is *always* lidless. Sizes run from the gallon down to the half gill, later adding the quarter gill; but since inception the smallest has been made in about a dozen different authorised and stamped sizes, from $\frac{1}{10}$ to $\frac{1}{32}$ pint. I even have one, made by soldering the upper section direct onto the base (i.e. no belly), which holds $\frac{1}{48}$ pint! Many of this series bear the Imperial stamp of 1826—and here, one would think, is a design new to the period. Yet once more one can look a few years earlier, for at least one specimen exists in O.E.W.S. which was abolished in 1826. It is, I suppose, possible that they were made for America (whose capacity standard remains O.E.W.S.). But we must remember that the old Ale Measure was, to all practical purposes, the same as Imperial; therefore it is probable that these were in fact produced before 1820. Michaelis has pointed out that a drawing by Hogarth entitled 'Gin Lane' published about 1760 shows a tavern sign depicting almost identically this form!

Since this type of measure is generally thought to be solely of Imperial Measure use we will take them in this period, with qualifications as to earlier manufacture. Michaelis was also the first to sort out their nineteenth-century dating. As usual, the same difficulties occur in getting an earliest or latest date on any piece, and any estimate of date must be taken as elastic. The verification stamps do not really help, as they could have been put on after manufacture. The maker seldom lived in the district where the stamp was applied. A measure might bear twenty verification stamps (some do) between 1826 and, say, 1900. That does not preclude its manufacture earlier than 1820. However, where 'GIV' or

Fig. 70
The bellied (or 'bulbous') measure which
is the one style most nearly confined to
the nineteenth century. Actually there is
no doubt that it was in use c. 1800, and
even now you may have a measure of gin
served in this style.

'WIV' are stamped, it can be taken that the measure or tankard
was stamped at latest before that monarch's death.

There are six main groups of bellied measures, and almost all
the bodies could have been made from the same mould (per size,
of course). One might for simplicity say that the handles are all
the same design between the two points of attachment. We then
have only three variables: (*a*) plain body; (*b*) rings round the body,
and where placed; (*c*) finial to handle. With so few variables, there
is little to learn! However, intensified and widespread research
may throw up fresh facts and modify dating.

Type 1. Plain body—no rings. Shapes do vary a little, and have not
settled down. Handle finial is a fairly well-formed ball.
c. 1790–c. 1830.

138

Fig. 71

The dating details of the very common nineteenth-century (and earlier) bellied measures. The handle finials and the rings are the only clues to dating, except that tall attenuated shapes are eighteenth century.

Type 2. Three or four incised rings above the lower point of handle attachment. Finial is a weaker drawn-out ball c. 1824– c. 1830.

Type 3. As Type 2, but with a slightly thickened band over the seam at maximum body diameter. c. 1826–50.

Type 4. As Type 3, but no incised rings on the upper curve. c. 1830–60.

Type 5. Rings on the upper curve only. Stouter handle with no finial—it fades away almost flush with the body. c. 1850 onwards.

Type 6. Has a more crude 'feel' and finish. Has a thick, heavy band round the seam, no other rings. 'Finial' is even less. c. 1880 onwards (largely Edwardian). (This was, and is, also made without the lip.)

The distinctive period details are less clearly definitive in the smallest size. These bellied, or 'bulbous', measures must have been used in very large quantities indeed—until a very few years ago they could be found everywhere. The gallon and half-gallon sizes were always scarce. The quarts have recently been bought for holding flowers, and it now needs considerable patience to make up a set of the six more usual sizes. The variety of types described adds considerably to the interest of the search for them.

Fig. 72
West Country measure, gill. No mark
Nineteenth century. Height to lip $3\frac{1}{4}''$.

The squat rather frog-like West Country designs are the only other nineteenth-century true measures. They are of two types somewhat alike: the West Country—short and squat, similar to copper measures—and the Bristol or 'Fothergill' type (after its principal maker). This is much slimmer and of far simpler lines; while the former are rare, the latter are very scarce indeed. They run from half gill up to very large sizes. Both these may have been made before Imperial Measure, too. The Irish 'Haystacks' are rather comparable in the essence of design. Reference to the chapter on Scottish and Irish will clarify the difference. One other type of curious measure is still to be met occasionally—the 'double egg cup', hour-glass shape. One cup is twice the capacity of the other. They are c. 1830.

In this period serious-purpose tankards can be taken as always lidless. Some—prizes for competitions, particularly rowing—bear lids, but only for prestige. These trophies of 1880–95 often have glass bottoms. It is fun trying to guess the reason for these glass bottoms. There are several suggestions. Take your choice. The Press Gang chat up a likely young recruit, slip a shilling in his tankard—and he is found to have accepted the King's Shilling. Since the glass bottoms started in about the 1780s, this may be true. Another explanation is to be able to see any hostile movement by your drinking companions. A third—to see if your beer is clear.

140

There is little to be of this century's tankards—most has already been said in the previous chapter. There are few designs which we can be certain were introduced after 1826 and it must be emphasised that the overwhelming majority of tankards to be seen are Victorian—but not all. Makers' marks are very inadequately recorded and much information can be added in this connection —names, districts and first-and-last dates. The alloy used is very tiresome for the cleaner. The oxide forms quickly, heavily, and it clings. It can be extremely tedious to do the job carefully and as one would wish, and there will be a temptation towards total

Fig. 73
Nineteenth-century debased wriggling, and linear engraving.

immersion. One must visualise the resulting lunar landscape before starting to strip—it will need a lot of handwork to gain a fine surface.

Tankards are very often inscribed with, not cursive engraving, but linear letters. They are a little difficult to read, but it is easier to do than the beautifully curved lettering of the previous century. It is particularly noteworthy that wriggling in a coarse hand was reintroduced—now only for the outline of the area of inscription. The linear engraving as to ownership is very often to be found under the base. This feature tends to be later rather than early. Tankards are quart to half pint in sizes.

The information on verification marks (in 'Marks'), as well as the eighteenth-century examples, brings the Victorian tankards to

141

a level of collectability. You, and I, and everyone else, have allowed these to trickle away without making our notes—or keeping examples. Now we can pay for our superiority or lack of foresight.

A subsection of tankards is beakers, or goblets (without handles). These are comparatively scarce now. It is quite possible that museums may possess some, although not having them on display, and examples sometimes appear in shops. We need evi-

Fig. 74
A considerable quantity of very similar beakers and footed cups were made in the second quarter of the nineteenth century. Footed cup, maker C. Bentley ('Cott. *O.P.*' 407), with 'hall marks'. c. 1830. Height 4⅝″. Stamped 'St P M W IV'.
(*Mr P. Starling*)

dence to date the slightly differing designs, on the lines of our eighteenth-century tankard chart. Almost all are outward tapering to a flared-out rim, and some have a waisted base. So it is a matter of dating the rings and bands. As far as I know, no consistent serious work has been done on these pleasant disregarded fellows. Footed cups, too, are sometimes to be found with two, one, or no handles. The latter (in pewter only) are mostly late eighteenth century, having either of the three likely handle and finial combinations. There are also three-handled cups, for passing from person

142

to person. Surprisingly, I have seen only half-pint sizes—one would expect at least quarts. Two-handled cups are used for ceremonial 'Loving Cups' at traditional functions. But would they ever have been in pewter? Perhaps they were decorative pieces commemorative of weddings.

Funnels probably go back much further than we have thought—they very seldom bear any mark. Some have bands of incised rings, others are plain. There is less than usual on which to build information. Inkwells, on the other hand, have preserved their youth as the most up-to-date class of Victoriana. Inkwells had been contained in the little pewter chests, or on the standish trays, in the previous century. In the nineteenth century the circular capstan, or loggerheads, design took over, subsequently with a wide base —10″–15″ approximately—to be found until very recently indeed in all Post Offices, and even now without the flange. What a curious place for the general public to say farewell to its use of pewter.

Casters, with various perforations, for sugar, pounce, salt and spices had abounded in the 1700s in a wide variety of shapes, and they continued in the nineteenth century. Their dating is comparatively easy by reference to similar styles in silver. They make attractive displays, particularly in breaking up a row of tankards.

Snuff boxes occur in pewter as well as in Britannia metal, and are often quite interesting. Sometimes hunting scenes are shown.

Spoons were still made—coarse kitchen spoons—and their period is easily recognisable. Manufacture of candlesticks probably continued unchanged from early in the century. Many other items are to be found, from whistles to syringes. Again, dating is difficult.

An enormous subject, almost completely destroyed or dissipated, is that of toy or doll's-house pewter. A prodigious quantity was made in the century under review, and is very fascinating. Some, but not all, is crudely made. Some, ornamented, is stamped out; some is neatly turned. The metal alloy used varies enormously from what seems like chewing gum to Britannia metal. I think it can be taken that the styles shown in doll's–house pewter were probably decades behind the current fashion.

Mantel ornaments, usually taken as early nineteenth century were certainly made in the 1760s, and examples may be earlier than expected. Tobacco boxes, tea caddies, tea pots, pap boats—all occur in the nineteenth-century range as well as the eighteenth.

A class which is perhaps outside our brief—medallions—does go back to the seventeenth century, but interesting specimens in

Fig. 75

Doll's-house pewter. An enormous amount of toy ware in widely varied alloy was made in the nineteenth century and earlier, and because of its poor quality and usage it has very largely disappeared. (Brothers love to melt lead!) No marks. c. 1860. Teapot 2″ diameter.

(Mrs E. Butler)

the nineteenth and eighteenth centuries can be obtained from numismatic firms.

Two things are worth remarking about the nineteenth century. First—that it is well worth inspecting each piece of your 'nineteenth' century pewter, and every piece you can see, for it may well be eighteenth century. One friend of mine decided early that seventeenth-century pewter was impossibly dear, and turned all his attention to collecting nineteenth century. But now it turns out that most of his acquisitions are of the previous century! Develop an appreciation of the nuances of manufacture, style and marks, and do not take Cotterell's *opinion* of nineteenth-century dating too conclusively. Yes—where there is proof (of maker's start or finish, for instance), but where it is his opinion—pit your own against his.

144

Fig. 76
Mantel ornament and tobacco jar of late
eighteenth century. The mantel ornament
is one of a set of four, each representing
a continent. This is Africa. Tobacco boxes
were made in many styles; the negro head
was an emblem of tobacco. No marks.
Height of ornament 4″.

(*Mr P. Starling*)

Fig. 77
Art Nouveau pewter. There is not very
much to be seen, and it has not yet caught
the serious collectors' eyes. Vase, c. 1900.
Hot-water jug, 1904. Maker 'Tudric'.
Height to lip $6\frac{1}{2}″$.

145

Second—there is a fertile field for research in dating styles, and extracting details from: beakers, footed cups, casters, candlesticks, and toy pewter.

Before leaving English pewter there are three more broad categories to be mentioned.

We have seen trophies in the 1880s to '90s. Now Art Nouveau burst on the scene, and extended its media to pewter items until c. 1905. Utterly scorned until now, they are 'in'—and probably will not fall.

Although dealt with later in conjunction with Fakes, we must be warned that many old styles—and old-looking styles—were 'reproduction' made—from 1925 to 1939 and after. These pre-war products are to be seen everywhere, and in some cases ignorant(?) dealers are putting very high prices on obvious reproductions.

It is always most difficult to write of the present, and few readers require it. There are a great many small manufacturers all over the country and one or two bigger names long connected with pewter history. Some of their products are replicas of old styles, some are akin. If you want to know what they are making, a visit to your local gift shop, jeweller or tobacconist may produce brochures and price lists.

9. Scottish, Welsh, Irish and Channel Isle Pewter

Introduction

It is very difficult to know how to give the noble Scottish pewter a fair proportion of attention, and in fact we cannot do so. Talk of Scottish pewter recalls tappit hens, and small newish lidded measures—perhaps not much else—but the subject has been its own worst enemy for collectors. Tin is not mined in Scotland, so pewter was expensive—very—and at that in a very poor country. So although it was made there from an early medieval period comparable to England, there was far less in the first place. Furthermore, its scrap value was not only high, but largely essential to new products. Political and religious changes caused much destruction of earlier pieces. There are other reasons, too, for its scarceness. Tappit hens—although there must be hundreds in existence—are such an epitome of Scottish character that they have always been admired, and have earned themselves homes which are reluctant to put them out, even at a price. Most other styles available are not nearly so significant, and a very great deal of flatware was not marked and is therefore unrecognised as Scottish, and some which is taken as English may in fact be Scottish.

147

Fig. 78
The tappit hen—the acme of Scottish pewter. Continental influence. Made in Scottish, Imperial, and local measures of capacity. This is the Scots pint, equal to three Imperial pints. No plouck. No mark. c. 1760. Height to lip 9".

(Mr W. Allen)

Successful Scottish emigrants living abroad probably feel a nationalistic yearn for symbolic company of their nation—what better than a tappit hen? So we shall have to pick a way through history, museum specimens and collections to give a balanced view. By far the best and most comprehensive work is L. Ingleby Wood's *Scottish Pewterware and Pewterers*. It is open to much criticism and could have been far better arranged, illustrated and captioned, and several classes of article were apparently not known to him. But it was published sixty years ago—about twenty years before Cotterell's *Old Pewter* . . . So all honour and credit to the man who undertook such an enormous amount of research in both product and history. Obviously much has been learned since 1910; collectors have exchanged their impressions, ideas and evidence with increased candour.

Pewter was made under the auspices of the Incorporation of the Hammermen, in various towns and cities. Because the population was both sparse and poor there was little demand and therefore, with other loosely kindred trades, they banded together in their Incorporation for protection and control. Those who wielded

148

a hammer within the Incorporation would probably include armourers, cutlers, blacksmiths, loriners, even goldsmiths in early days, clockmakers, bell founders, plumbers and tinkers. But each craftsman was supposed to stick to his own craft, and to deal only through the recognised channels. The Dean (Head) of the Guild was the President of the Chamber of Commerce, of the Masters' Federation and of all the Trade Unions, as it were, all rolled into one. He was also the Weights and Measures Inspector.

It is most important to appreciate the very strongly different background to the Scottish culture and nation in the sixteenth, seventeenth and eighteenth centuries. For long Scotland and France, and Holland, had found it politically wise to stand together against perfidious Albion. There had been more cultural exchange between Scotland and France than with England. Research into etymology and many outlets of art and culture confirm this. The result in our subject is that little true originality of design was apparent, evolution from other media not uncommon, but adaptation of nationalistic traits produced some superb examples (e.g. tappit hens), and some much less happy. The national sturdiness is *always* made apparent, without introducing the grosser elements of the Georgian influence. One can say that almost invariably Scottish designs of pre-1700 are acceptances, adaptations or modifications of Dutch and French, and, after the Act of Union 1707, of English.

We southerners perhaps think of the earlier Scots as rebellious fellows. In our field they were saints personified, by comparison with the spiv and 'mod' fellows in London in the seventeenth and eighteenth centuries. Quality of metal and workmanship was far better observed or controlled. The pity is that so little is in circulation now, although a few years ago late eighteenth- and early nineteenth Scottish pewter was too common to be of interest. It probably still lurks hidden away in many a Scottish home. Only the week before writing this I came across a good mutchkin-size tappit hen very unexpectedly.

Marks

On marks, there is surprisingly little to say which is worth taking space in a book of this size. You will not find marks such as those on the Edinburgh Touch-Plate (actually two plates—but the second only bears two marks). Much Scottish pewter prior to, say, 1745 bears no mark. Marks on Scottish pewter are normally helpfully obvious—an undeniable flavour of the name, or town, being mentioned, and latterly, in the nineteenth century, makers' names cast in a large circle, all combine to make the elementary stage of studying Scottish marks rather mundane. Although so much of the Scottish aspect differs greatly from English, verification marks of capacity usually follow the same rules as in England and therefore would have meant duplication if repeated here. 'Cott. O.P.' takes in all the Scottish marks known to him and also gives illustrations of the types.

It was not until after c. 1750 that the Scottish pewterers started to use the English type of mark, very often a rose, and name. 'Hall marks' were not adopted until about the Act of Union (1707). The patriotic nationalistic thistle appears almost invariably in one shield of the four 'hall marks' and often the rose in another. Add a contraction of the town in a third and one has only the fourth for maker's initials. However, the permutations are quite adequate for identification. For those really interested in minute historical detail, and names of pewterers in various towns —a local library would probably be able to get Ingleby Wood's 'Scottish Pewterware'.

The crowned X was used faithfully and more consistently accurately than by the unruly English pewterers: it came into use some thirty years after England. It is curious that whereas in England in the seventeenth century the touch was always on the back of the rim, in Scotland it was on the front—but so it was in England very occasionally in the sixteenth century. One may come across pieces with a puzzling town-arms mark and initials. This—seldom found now—is on measures, and is the arms of the Dean of Guild's town and his initials. It is his 'verification' mark. Nineteenth-century verification we will deal with under the section on Measures. In the nineteenth century makers' marks were 'industrial-revolutionised' and were cast in the bases or lids of pieces.

One touch mark which until quite recently could be seen frequently—and which will undoubtedly still turn up at sales and specialised dealers (and I expect in the States)—is square, with a ship in full sail, and captions surrounding it such as 'Success to the British Colonies, Maxwell', 'May the United States of America Flourish. S. Maxwell', and 'Success to the United States of America. Maxwell'. Stephen Maxwell of Glasgow, c. 1784–c. 1820, must have furnished a lot of households and churches then, and collections now, with flagons and basins.

Ecclesiastical

With a race so fervently religious and so thirsty it is difficult to know with which main category to start—Church or measures, but we will let the Church come first since we have just mentioned church items.

Just as in England, and in fact all Europe, monks were buried with their pewter chalice, denoting their calling. Some of these chalices have been unearthed. Medieval pewter of any type is almost non-existent. Of pewter itself, it appears to have been made in the fifteenth century, but it was not until 1493 that the pewterers of Edinburgh were accepted as a craft into the city's Incorporation of Hammermen. It is a direct contrast with England and Holland, but in harmony with France, that at no time was any incised (engraved or wriggled) decoration added to pieces. One bowl with punched decoration somewhat similar to English sixteenth-century and seventeenth-century dishes is in existence, its function unknown—but neither Church, Kirk nor household had any use for decoration (save I.H.S. in a glory).

It will be helpful, if I give a brief background (perhaps too brief for some) of the religious troubles of the period. The Reformation in 1560, which put the ardent and destructive Presbyterians in control, resulted in all popish church furnishings being melted down. Back came the Episcopalians in 1617 until 1638, reintroducing the chalice and paten, for example. The fanatical Covenanters melted all before them in 1638, but the Episcopalians came again in 1660–88. Although disestablished, they gathered favour and by 1745 had regained a footing. The rebellion of that year sacked all their places of worship—but yet again they gradually regrouped. With all this religious see-saw, vessels were sometimes used by the succeeding sect, sometimes replaced. At one time the kneeling of the Episcopalian communion gave way to the Presbyterian custom of being seated round the table. Ritual affected the size of the cup. Infrequency of celebration meant at times huge congregations—neighbouring villages would come to partake, and so large Communion services of even twelve cups and six flagons were used. To simplify it, let us say that the whole picture is most confused. When it is remembered that the Church and the Kirk, like the populace, were poor, one understands that cups were bought, of domestic style, 'off the peg'. Styles, too, altered little, due to using the same moulds. Chalices in the late seventeenth and early eighteenth centuries were

151

Fig. 79
Scottish Communion cup and flagon. The
cup is engraved 'Relief Church Leith 1822'.
The flagon is engraved 'Belonging to the
Associate Congregation of East Barne
1781'. Maker John Gargner, of Edin-
burgh ('Cott. *O.P.*' 1808). Height to
lip 9″. It is interesting to note that both
'Relief Church' and 'Associate Congrega-
tion' are disestablished groups from the
Church of Scotland.
(*Mr W. Buckell*)

often what we would call tumblers, based on the Dutch beakers.
But the movement after 1745 meant buying new services (because
all had been destroyed) and much of this plate still exists, mostly
in its spiritual home—but by no means all.

The initiation sacrament of a baby into its religion needs
explaining as far as pewter vessels are affected. The vessels
are the laver, for water, and the basin. Usually of pewter, the

laver was a smaller version of the flagon, with a spout which was occasionally constricted to a narrow channel. It was used to carry water to fill either the font or the basin, or to be used by the clergyman to pour the water direct over the child's face. Both laver and basin were probably introduced in 1617, after most of the fonts had been removed at the Reformation. Unlike flagons, lavers are sometimes lidless. The form, as so often is the case with Scottish pewter, was the same for both domestic and religious purposes. Both sects used both vessels in the different ways mentioned and sometimes one sect adopted the other's ritual.

Fig. 80
Laver—for baptismal water. This may have served double duty, as a flagon as well as laver. Maker Adam Anderson ('Cott. *O.P.*' 75, Edinburgh Touch-Plate 134). c. 1770. Height to lip 7¾".

The basin changed remarkably little from 1620 to 1800 and is really most unexciting, being similar to the English form, with mouldings on the narrow rim and with steep sides. A smaller footed bowl with sloping sides took over, but, with its small base, must have been very unstable. Some—as with flagons and chalices of Episcopalian use—bear 'I.H.S.' within a glory. Basins had other useful functions—for receiving the collection in the porch and also for communicants to deposit their tokens. Plates, both large and deep (14″ × 1½″), were also used for collections.

Tokens were necessary for admission to Communion, and were earned by passing searching catechisms as to knowledge and character carried out by the elders, who then distributed the token to those who were satisfactory. Communion was celebrated infrequently and neighbouring villages as well as parishioners would seek tokens. Numbers distributed were counted and the names checked off against a list as the communicants put their tokens in the basin. The tokens are of lead, or pewter, were formerly of 'hammered' manufacture and, in the nineteenth century, cast (Plate 13). Although completely different in use to coinage, communion tokens are usually a numismatic subject and the bigger dealers usually have some available.

We have led round to Communion, the three vessels being, of course, the cup or chalice (depending on sect), the paten or plate, and the flagon.

Communion cups are in circulation occasionally. The earliest were probably the tumbler type previously referred to. Then the cup was raised on a short stem, later elongated, the stem bearing bulbous drops and balusters somewhat indiscriminately in period. It is an oversimplification to say that the later the cup, the taller and slimmer the cup part, and the less apparent any stem swellings, because some late cups are very wide. Marks do not help at all, for they are absent.

Patens were probably the same as, or very similar to, domestic plates. Like other pieces of communion plate they are often, particularly latterly, decorated with 'I.H.S.' and a glory. The broad-rim style of England appears to be rather unaccountably missing, but the following triple-reed and narrow-rim types are represented, the latter having straight sloping sides. A special type would seem to fit in here—a narrow wavy-edged type on three feet. After this, English domestic types are followed—single reed and plain rim. The former sometimes bore beading round the edge. Unlike the English Church, the plates for the bread are sometimes very large—up to 20″. Eighteenth-century plates found in England usually have marks—not always identifiable—but this does not follow in Scotland, for it is unusual to find marks on pieces prior to c. 1750. In the seventeenth century there are proportionately more plates to be found without marks, and there is little evidence to link them with Scotland—just the possibility.

No-one knows, has suggested, or perhaps even wondered where the typical Scottish flagon was first made. Obvious, one might think, but it is not generally realised that the same style was used in England in the c. 1710–c. 1730 period. It is most unlikely that the very many widespread parishes in Southern England would have adopted a Scottish native style.

On the other hand, the type bears no relation to, and takes no part in, the evolution and development of English flagons. In

Fig. 81
Very fine engraved Scottish flagon and
dish. This is by far the most common type
of Scottish flagon, which originated in
England in the early eighteenth century.
In Scotland it continued well into the
nineteenth century. Flagon by William
Eddon ('Cott. *O.P.*' 1503). Height to lip
11″. Dish by Richard Grunwin ('Cott.
O.P.' 2040). Diameter 16⅜″.

(Collection of the late Mr E. Hunter)

England they came and went quite quickly, seldom if ever engraved, seldom by identified makers. In Scotland they appeared
early in the eighteenth century—and continued in the nineteenth
century unaltered—exactly the same body, thumbpiece, lid,
handle and terminal. I think (and this is observation rather than
applied testing) that the earlier lids are rather fuller of curve than
the later—which does not seem logical—and the lower handle
attachment is flush early, and on a strut later. One would have
thought the lid curvature would remain constant, for were the
makers not really making their moulds pay ? They used the same
moulds for a century. Another clue as to dating can be gleaned

155

from the styling of any touch mark to be seen (in the middle of the base inside) and 'hall marks' on the lid or drum. Very occasionally there are projections on the front of the lid as in dome-lidded tankards—quite small and pointed—indicative, of course, of the earliest. Spouts, too, appear—often very crudely—and probably some of these were put on long after 'birth'—so in itself a spout indicates lateness—but not necessarily of the body. Parallel, and occasionally interspersed, are the other South of England types (NOT York or Lancashire types) with tapering drum and dome lid, and subsequently the 'spire' type with 'broken' handle—but these are rare, and are invariably made in London. The accepted flattish lid type is the only Scottish flagon likely to be met, and is not infrequently inscribed. Otherwise, identification is by makers and type of marks. Dating is by presence or absence of lid projection, or lid attachment. Place of origin is very difficult without much help from the piece itself.

Domestic

Turning to domestic ware, we have already touched on plates, and noted that plates of the sixteenth century are exceptionally rare—anywhere. Scottish seventeenth-century plates, too, are much more rare than English. Probably there are many deeper plates to take the broth of the nation than was customary in England. Where are the broad-rim plates, narrow-rim, and triple-reed? Where are all the Edinburgh Castle touch marks, other than on the touch-plate? Let us hope that there are many somewhere, unrecognised.

Three other most attractive ranges of English pewter are also a void—candlesticks, salts and spoons. No doubt the same reason is largely responsible for our disappointment—strict control of damaged or unwanted pieces—they went back into the pot, and native prudence saw to it that spoons were not entirely lost.

So let us move on to the glory of Scottish pewter, as we know it now—the measures. Probably most people who are interested already know that Scottish Measure differed considerably in capacity and name. In fact there was endless trouble in trying to establish and enforce a standard in which official measures themselves varied considerably. Although there are some 'odd' sizes, the position is quite simple, at least theoretically. The earlier measures were apparently in three sizes only—Scots Pint, Chopin, and Mutchkin. For simplicity, since English Ale standards were so close to Imperial, we will compare with the latter—NOT with Old English Wine Measure.

Scots Pint	= 3 Imperial Pints	= 60 fl. oz.	Qt. = 40 fl. oz.
Chopin	= $1\frac{1}{2}$,, ,,	= 30 fl. oz.	Pt. = 20 fl. oz.
Mutchkin	= $\frac{3}{4}$,, ,,	= 15 fl. oz.	$\frac{1}{2}$ Pt. = 10 fl. oz.
			Gill = 5 fl. oz.

Later, successive smaller sizes, each a half of that above, were made.

When Imperial Measure was put into practice in 1826, Imperial sizes were made in the current measures—the famous tappit hens (Plate 14). But the old Scots Measure died hard, and continued to be used until 1855, when it was banned completely, and the Imperial Measure equivalent should have been stamped on them, e.g. 3.I.G. (Imperial Gills). Many examples stamped with equivalent fractions exist, such as $\frac{3}{4}$ I.S. There are also some odd and local, now rare, capacities, such as 'two glass', 'four glass'. The larger sizes usually carried a blob, or pimple—the plouck— on the inside of the neck, and denoted the full measure mark. In its absence they should be measured to the lip.

The range of styles is covered by 'Potbelly', 'Tappit Hen', Balusters, Lidded Bellied measures, and Thistles. Most have variations. Before going further I invite you to look at the profile view of the inside curve of the handle of any and every illustration of these measures that you can find, except the thistle. This point does not appear elsewhere in print, but is most useful—in every Scottish measure you find an extra long 'straight' under the top of the handle. This is a remarkable feature which lasted from c. 1680 to c. 1880. It occurs also in only exceptionally early English measures, and in two very rare types of Scottish capacity which *may* be North Country English. It defeats my deductive powers why this peculiar detail of the handle should have occurred at all, why it was adopted by the Scottish pewterers, and why it lasted so long.

It is indeed strange that Ingleby Wood, despite apparently a wide knowledge of Scottish pewter, makes no reference to, and does not illustrate, the well-known potbelly type. Examples are generally taken to be the earliest of existing Scottish measures. Probably existing potbellies are older than existing tappit hens. But one of the latter is sculpted in a statue on the fountain at Linlithgow, which the Ministry of Public Buildings and Works

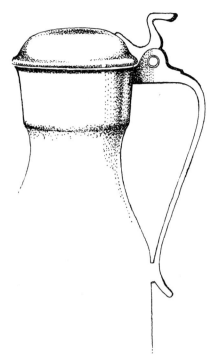

Fig. 82

A most important point to appreciate—
the long straight section under the top of
the handle on *all* measures except the
thistle. (It does occur on very early
English balusters of fifteenth to sixteenth
century, even to early seventeenth century.
It *may* also have been used in some of
eighteenth century from the north of
England.)

Fig. 83

The potbelly—the predecessor of the
tappit hen, also obviously of Continental
descent. They date from c. 1680 to c. 1740,
latterly in a slightly modified lid form. In
this example the capacity is 66 fl. oz. to the
plouck (not conforming to any standard),
yet there is a mark on the handle, a portcullis within a thistle, and 'S.I.', which
looks too crude to be a maker's mark,
and may be some type of verification. No
other mark. c. 1720. Height to lip 8½".

(*Mr W. Allen*)

(Scotland) confirms is sixteenth-century work. I have not seen it, but Ingleby Wood's illustration of it shows unmistakably a tappit hen, although the bottom part is shorter than those we know. It is undeniably of Netherland influence, as indeed is the potbelly. English pewter has no counterpart with either. The potbelly seems to run longer than most people imagine—very roughly from c. 1680 to c. 1740—and the later lids are lower, with narrow flange. The handle, too, is weaker in design. Potbellies are rare—certainly rarer in circulation—and some have made up replaced lids.

To return to the extremely attractive and dignified tappit hen, which is everyone's idea of the acme of Scottish pewter. Curiously, the Pint is very masculine, the Chopin dainty and feminine (husband and wife) and the Mutchkin reminiscent of a chunky boy in shorts. As well as the plain-lidded variety, it was made latterly (early nineteenth century) with a knop on the lid, in the main three sizes only; and also in a lidless form. (The same mould for handles was used for lidded and lidless—the hinge boss on handles of the lidless being blind.) The earliest evidence, after the sixteenth-century fountain, is one bearing a dated touch of 1669, which was probably made during the seventeenth century. They continued to be made up to c. 1850–60. (This type was also made in a modified style in brass. I know of five examples, all Scots Pint approx.) The total range made of known different sizes including all Scottish, local, and Imperial capacities is said to run to about nineteen. Be very wary of lightweight specimens with an X, and all or some of Bush and Perkins 'hall marks'. I know, for in my early days I bought one of these reproductions. I do not suppose Bush and Perkins had any reason to make 'real' tappit hens!

Before the balusters, it is just conceivably logical to turn to the rare thistle measures. These are late—probably latter half of the nineteenth century—and were very soon ordered to be destroyed when it was found that they would not empty their contents when tipped to the statutory 120° laid down by the Act of 1907 (thus retaining some spirit for the landlord). There are very few in circulation, and I am not sure whether reproductions or fakes of such a late product exist. It is hard to tell. The reason for inclusion at this stage is to suggest that their design is based on the potbelly/tappit hen, and is their descendant. Note that the handle is an odd man out in style.

There are three types of Scottish balusters, which are the latest to have been made in any country, and are almost all post-1835.

Fig. 84

Thistle measure. These were made very late, and were subsequently found not to conform with the Act of 1907 which required contents to be emptied when tilted to 120°. They were ordered to be destroyed, and are now therefore very rare. Note the handle, of which the finial is missing. It should sweep away below the strut. Mark indecipherable. Latter half nineteenth century. Height 2″.

They consist of the Ball (and bar), Embryo Shell, and Shovel. The former is distinguished from the early English ball by the *bar,* instead of an inclined attachment for the thumbpiece. (However, the rare English variety of c. 1690–c. 1740 has a very similar bar.) The belly of these late Scottish balusters is very high by

Plate 13 Communion tokens. (a) NBK 1714 (*Newbattle Kirk*). (b) Kennoway 1835. The 18th century tokens appear all to be of hammered manufacture, and the 19th century, cast.

Plate 14 Crested tappit hen. Chopin size, plouk in neck. No mark. Early 19th century.
8⅛″ to lip. (Dr E. Oastler)

Fig. 85

(a) Scottish ball baluster, half pint Imperial, stamped G IV $\begin{smallmatrix} R & W \\ D & G \end{smallmatrix}$ (probably R. W. Dean of Guild). (b) Embryo Shell, half pint, stamped W IV, V.R., and E.R. Height to lip 4½". Both by Robert Whyte, of Edinburgh. c. 1830. The Shovel thumbpiece is more pointed at the top, and is nipped in at the waist.

comparison with all other balusters, and the lower handle attachment is quite a long strut. The handles, of course, have the long straight under the hinge. An outstanding feature of all the Scottish balusters (and some rare ones which may be North Country English) is the anti-wobble flange under the lid to fit within the rim, _to prevent lateral leverage on the hinge lug. Curiously enough, this feature, like the handle straight, is also present in some very early English balusters. This ball type is rare in Scottish

161

standard (before 1835) but was recently quite common in Imperial Measure. It was also made lidless, and the range is usually half pint downwards, in four sizes. The embryo shell is so called for a very weak design of thumbpiece rather like an unribbed half of a seashore shell. It bears all the other features of the ball, but I do not know if it was made in Scottish Measure. The shovel is more rare, and is probably the youngest of the trio. Note that the maker's mark in these and some late tappit hens is cast in the

Fig. 86
All Scottish baluster lids have a centring device—an anti-wobble rim, to save wear on the hinge lug.

base, not struck. These three types were made by, amongst others, James Moyse, probably as late as 1870.

Some earlier rare types may possibly be met, and if I do not mention them you may doubt their authenticity. There are very few (I know of four) slim ball measures which I date very loosely by their lines (and I would welcome correction) at c. 1680 (Plate 15). These have the straight and anti-wobble rim. They are in Scottish Measure, but *may* be North Country English. There are also a very few English Buds made in Scottish Measure. This is a most interesting fact, and two I possess are, I suppose, c. 1680—

162

and most surprisingly, these may be the earliest existing Scottish measures. There is a curious type which, but for the straight and the lid rim, I would put as English, being near Old English Wine Measure. This is the Spray, which is only a little less rare than the slim ball. It is obviously of mid-eighteenth century judging by

Fig. 87
Glasgow single-dome measure—direct descendant of the baluster. The double-dome variety is now rare in circulation, despite the late date—c. 1840. Height to lip $3\frac{1}{2}''$. Edinburgh measure. Note the concave curve of the lid, and obvious derivation from the bellied measures. Stamped V.R.479 (Midlothian County). Maker J. Moyse (cast in the lid). c. 1880. Height to lip $2\frac{7}{8}''$.

its lines (Plate 15). I have seen it only in pint and quart. These, too, may be Newcastle-made. None has an identifiable touch.

So to the pleasant lidded Pear measures of three lid types— Edinburgh, Glasgow double dome and Glasgow single dome. The Glasgow measures, like a dome-lidded baluster, are taller and slimmer than the Edinburgh, which are rather bulging and whose body is similar to the English bellied measures. The former are of a much happier silhouette as a development of the baluster than

163

the bellied measures. These last varieties of Scottish measures bear the maker's name cast in the underside of the lid.

The verification marks on nineteenth-century measures follow roughly the same pattern as in England. In 1826 each authority had a stamp with its emblem and the monarch's initial and number, crowned (e.g. G.IV. or V.R.). After 1878 the 'heraldry' was abandoned for a district number, again with the reigning monarch's initial crowned. A few authorities did not adopt the uniform design of 1878, which may be confusing. As in England, if one wishes to trace the origins of a stamp, prior to 1878, the local Weights and Measures Inspector will be pleased to help, otherwise the list of numbers published at the end of the chapter on Marks will be useful.

Wales

Most regrettably, I can find no evidence whatsoever of pewter manufacture in Wales. For such a strongly nationalistic country this is an extraordinary omission, yet there has been very little discussion even on the subject.

There is no evidence on manufacture, makers or types. Here is a challenge to someone. My own contributions are scant, and not very helpful generally. In chronological order—first, there is the truly fantastic bowl (referred to in the chapter 'Romano-British to 1600'). It is presumably a font bowl, with Celtic or Anglo-Saxon decoration. It came from Wales, and that is the only weak evidence for its inclusion here. Efforts to find parallels for the decoration have been fruitless, and it will be recalled that the decoration is carried out in the chisel-punched (or chip-carved) method, which dates back to the first millennium A.D., and is not known on medieval pewter.

A fine flagon is portrayed, which is not in the run of English styles, and is engraved 'Thomas Owen. Clr. Moses Roberts. E. Ellis. Wardens. 1764'. This piece, to judge by lid, thumbpiece and body, is c. 1710. The mark is illegible save '..ONDO..' (part of 'London', which does NOT prove London manufacture). The names are, all three, a very strong clue to its provenance, but it is perhaps wishful to think we have found a purely Welsh style.

The third is not evidence, only custom—in Wales, and I think probably just over the Marches, it was the custom to stand plates face to the wall on the dresser, and to clean the exposed underside. Constant recurrence of a maker of bright-bottomed plates could suggest that he was a Welsh maker.

164

Fig. 88
Is this an exclusively Welsh flagon? Engraved 'Thomas Owen Clr. Moses Roberts E Ellis Wardens 1764'. Strong evidence of Welsh use. I do not know of a similar flagon. The features—lid, thumbpiece, handle, finial, and body—all indicate c. 1715. Mark, only '. . ONDO . .' (LONDON) is visible. Height to lip 10¼".

Irish

Like Scotland, Ireland was poor, and certainly could not support
a Company. Pewterers worked within their Guild of Hammermen.

There are a very few references to medieval pewter, and several
to that of the seventeenth century. But early Irish pewter appears
to be non-existent. The fact that I do not know of any much before
1700 may be gross ignorance, and I will welcome correction (on
this or any other point mentioned anywhere). Cotterell shows two
chalices of the late seventeenth century. There are some church

Fig. 89
Irish chalice and flagon. Chalice, no mark,
c. 1765. Height 8″. Flagon, 'hall marks'
of John Heaney, of Dublin ('Cott. *O.P.*'
2242). C. 1785. Height to lip $9\frac{1}{8}″$. Note
the coiled-spring appearance of the
handle.

(*Mr W. Allen*)

166

flagons in body and lid forms similar to the Beefeater, but the features are mixed. The handle is a wide 'S', like a spring, with fishtail terminal, and the body carries a spout: one feature of c. 1670, one of c. 1710 and one of c. 1760. Add a period for migration and adoption of style, and the computer says c.1725. This body shape, with extremely wide base and the same handle, is carried on through the century, a dome lid supplanting the previous lid. The eighteenth-century chalices, too, are easily identified, having nearly parallel sides of a rather narrow cup, and very stout stems.

There appears to be no evidence at all of measures, plates and other ware even in the eighteenth century, which is extraordinary. However, in the early nineteenth century, we run into the really

Fig. 90
The Irish Haystack, and noggin, measures. Very attractive styles, originating in the first half of the nineteenth century, still reproduced today. A 'noggin' is also a synonym for 'gill'. Haystack by 'Munster Iron Company', of Cork. Height $4\frac{3}{4}''$.

(*Mr P. Starling*)

delightful and distinctive Haystack measures—probably made for Imperial Measure in 1826. Note that they are still made today! These ran from one gallon to half noggin (a noggin being equal to a gill). There is a strong similarity in line to the West Country measures. Finally there is the series, again being reproduced today, called 'Noggins'. They are obviously sired by the baluster, but have neither lid nor handles, and run from half pint to quarter noggin; the latter size has concave sides. I once owned one of an earlier variety, but now regret having passed it on. It was of half pint, taller, slimmer neck but wide lip, and the belly was carried

Fig. 91
Irish spirit cups—probably mid-nineteenth century. 'Hall marks', 1. T.C. 2. Griffon to dexter. 3. Harp. 4. $\begin{smallmatrix} \times\times\times \\ \times \end{smallmatrix}$ (these are incised, not standing in relief). Height $2\frac{1}{2}''$.

rather lower. It had a definite air of bud, and was, I suppose, late eighteenth century. Small spirit cups were made and carry the maker's 'hall marks' showing the Irish harp.

Channel Isles

The pewter of the two main islands is in a curious situation. The isles are much nearer France, and the styling of their measures leans heavily towards that country. Where the makers are identified, they are very strongly localised Channel Isle surnames; nearly all add 'Made in London' to their marks, and some are known to have had London addresses.

There is little range, there being two special types of measure,

168

Plate 15 Slim Scottish Ball, and a Spray, balusters. Both types have the "straight", and anti-wobble rim. (a) Quarter mutchkin. c.1700? (b) Eighth mutchkin. "I.H." in heart. c.1690? (c) Spray, 16 fl. oz. c.1720. Ht. to lip 5½"

Plate 16   Britannia metal Mace. These were used by the Friendly Societies when they were creating ritual to foster allegiance. The maces have several different emblems, such as Sun, Moon, etc.

Fig. 92
The renowned Channel Isle measures.
(a) Jersey. Note plain body, Normandy-like shape, almost no flare-out at the base. Maker I. de St Croix (struck under the lid) ('Cott. *O.P.*' 1360, *L.T.P.* 833). c. 1760. Height to lip $6\frac{7}{8}''$. (b) Guernsey. In this example, rings round lip, neck, and belly. No mark. c. 1800. Height to lip $7\frac{1}{2}''$.

(*Mr W. Allen*)

and very ordinary English-type eighteenth-century plates. Of the measures, a large number are not marked. There are two similar but very recognisably different styles for each of the two main islands.

The Guernsey type has a well-defined foot, and definite baluster-like curvature, except that it does not swell very much out of the neck and has no collar. They have incised rings round the neck and belly. Both types have heart-shaped lids, and twin-acorn thumbpieces as in Normandy. They are late eighteenth century to mid-nineteenth, and run in three sizes—three, two and one pint respectively.

The Jersey type is not so refined, and is even more Normandy-like, having far less definite curvature and almost no nip into the insignificant base. There are no bands of decoration. The range is large—from half gallon to very small ones. The capacity is very confusing, even chaotic, appearing to take in local vagaries, metric and Imperial. Latterly this type was also made lidless, but generally appears to be earlier than the Guernsey type, judging from the likely working period of makers whose names are known.

10. Britannia Metal

If you define 'pewter' as an alloy of tin, you accept that the tin was alloyed with (a) lead and (b) copper. Those lucky enough to own copper-alloyed examples would not take happily to any suggestion that their fine medieval plates are not pewter. What then if we logically include another class—(c) tin alloyed with antimony? The pewter collector hitherto has not wanted to know it or be associated with it. For this is Britannia metal. If his selectivity were based solely on lack of space, it would be understandable. But if it is purely a matter of age—well, pieces exist two hundred years old, rare and collectable pieces, desirable for design and craftsmanship too. Pewter circles must soon come to terms with it!

This book is dealing with collecting *now,* and in *the future,* and so Britannia metal is welcomed into the fold. This is not to open the flood-gates, and suggest that any collector should therefore buy any Britannia metal he meets. No pewter collector does so with lead alloy. Knowledge, taste and discrimination play a large part.

Most unfortunately, Cotterell, the most prolific writer, and therefore the generally accepted greatest authority on pewter, lost no opportunity to decry Britannia metal, and told one how to weed it out to get rid of it. Thus the very people who had a taste and eye for Britannia metal were told they were beyond the pale if they possessed it, and until recently it has been below the dignity of most collectors to find even lumber-room space for it. Our American friends have been far broader-minded, for the simple reason, if no other, that it plays a very large part in their tin-alloy ware. They have also had the advantage of being less influenced by one insular writer who was looking further back into history. Cotterell talks of 'nightmare', 'avoid...', 'never buy...'. He may have had a point then. When he wrote, there was a large supply of old pewter available at reasonable prices, all of it older than Britannia metal. It is the reverse now. Why 'bother' with it now? Britannia metal is in almost the same position as nineteenth-century pub tankards held forty years ago. They were so common that they were disregarded. Yet here is a class whose earliest specimens go back 200 years, and have much to commend of the Industrial Revolution. Until very recently most Britannia-metal products were only of any value at all as scrap metal. Think what has happened to the whole antique trade in the last ten to twenty years. So let us start to look forward in collecting.

It will be recalled that throughout its manufacture pewter was usually trying to look like silver. The Industrial Revolution released ingenuity of mechanism, chemistry and thought. This was the period of enterprise and emergence. The first we know, as early as 1769, is of James Vickers who produced 'Vickers White Metal', which was 90 per cent tin, about 8 per cent antimony, plus copper and bismuth. He soon dropped the last two ingredients. He had evolved a new medium, not an improved pewter. It was far stronger and harder. It could be made much thinner (saving metal), but, much more important, it could copy the intricacies of silverware much more truly. And it was made the same way—spun. Pewter was always cast, but Vickers metal was put together from sheets. At first the purer silver lines were copied, and early Britannia-metal wares are gems of delicate lines, far removed from pewter's heavy simplicity.

It was not very long before the commercial attractions of the mass market were apparent, and by the mid-nineteenth century almost every conceivable form of additional design and decoration was thrown together on pieces, almost comparable with a grossly overdecorated fun-fair stall. Class gave way to mass. This new alloy was very extensively used as a base for electroplating too, and the majority of what we regard as ghastly gewgaws so frequently seen (the horrific teapots) underneath will be stamped 'E.P.B.M.' (Electroplated Britannia Metal)—post-1850.

Fig. 93
A really delightful example of Britannia metal. In its early years the best simple silver was faithfully copied.

(*Mr A. Cotman*)

The vast amount of decoration was applied by power presses, and usually there is only one seam made after the spinning. It is visible on the inside, running vertically (contrary to the circular seam we are accustomed to in pewter). The odd parts—handles, knobs, feet, spouts, etc.—were cast of a hard pewter.

The alloy is a brute to clean. The oxide is exceptionally hard, and the metal, being so hard, shows scratches from anything but the finest emery. It may be necessary to bring in the aid of hydrochloric acid or caustic soda—but the greatest aid is patience allied to persistence. All stages take longer, but the eventual result is superb, although bright by pewter standards.

Fig. 94
Britannia-metal makers always stamped the company name in full, and used no emblem. It was stamped incuse, as opposed to the legend in relief in normal pewter touches. Style numbers nearly always accompany the name. Sheffield was the major centre.

Fig. 95
Range of very well-designed, well-made, and desirable Britannia metal.

To identify this alloy is usually very easy indeed—if it is marked. The makers stamped their name and catalogue numbers on the base, the letters and figures biting in, which is contrary to pewter marks, which leave the design standing in relief, as on a coin.

Fig. 96
Although not so overdecorated as some late Britannia-metal teapots, surely by any standards this is a horror of design.

(However, in the nineteenth century some makers of pewter—e.g. James Yates—stamped their names on tankards, etc., in the Britannia-metal method). But the harder metal, the catalogue numbers, and the styles themselves (very seldom the same as pewter) clearly proclaim the alloy. Just occasionally, as I had recently, one gets comment from fond possessors of 'very early pewter—it has the date 1397 stamped on it'. Alas, the florid teapot is c. 1860, and we can decipher 'owner's initials'—E.P.B.M. and the catalogue number 1397. I know of a charming cream jug with the figure 4 underneath—certainly not A.D. 4.

Of the makers, those well known or likely to be encountered from 1770 to 1833 include James Vickers, Dixon and Smith, J. Wolstenholme, P. Ashberry, J. Dixon and Son, J. Dixon and Sons. The latter *did* make pewter as well—lots of it, despite Cotterell's investigation. Many nineteenth-century bellied lead pewter measures bear their stamp. [Since writing, Mrs Nancy Evans has published in America, a list of some 124 Britannia-metal craftsmen working in Sheffield prior to 1861.]

Items made include: teapots, cream jugs, coffee pots, church flagons and cups, casters, mustard pots, salts, plates, candlesticks, lamps, urns, dish covers, spittoons, maces (Plate 16), tobacco jars, and multitudinous other items.

Most Britannia metal except the latest and worst has been thrown away. Someone both patient and assiduous could still put together a small and very pleasing collection at little cost. But you will have to plough through many hundredweights of overdecorated teapots. Perhaps another generation will have a different eye for beauty—and these very gewgaws will be much sought after. Such is the way with antiques. But I have yet to hear of a genuine interested research collector whose medium is Britannia metal. This statement may be the fuse for a bomb of protests—but in the meantime I would say that if you feel any sort of affinity for Britannia-metal 'pewter', get busy and form a collection. It will be a nice exercise of taste, patience, search and research.

11. Fakes and Reproductions

When interest and literature in pewter developed strongly in the 1920s and '30s, some of the keen collectors let it be known what they particularly wanted. It was remarkable how quickly they were offered great rarities. There was no wariness, there was no criticism, either of their own or of others' pieces—there were no suspicions. These fakes, for such they were, were beautifully made technically—details of manufacture were beyond reproach; line, type, marks, and appearance of age all accepted. It is not known how many different sources there were (or are), but a great many were made. I do not suppose many have been destroyed, so they are virtually all in existence in England and in America. A good pewterer who knew the technical detail could cast them quite inexpensively in plaster moulds taken from originals. It is very easy to be taken in by clever fakes, for on seeing a rarity available, one's desire wells up, suppressing discrimination and discretion. All collectors have had them, and not only in their earliest days. Price is no criterion. At a major sale pieces may be detected by the majority—they may probably not fetch the full market value of true pieces. But they are to be seen fully priced in shops probably unrecognised by the vendor. Having been made some forty years ago they may well have acquired some true dulling oxide. They were treated, battered, repaired, and are at their 'best' when dusty and dirty, and, complete with dents, scratches, nicks, tears, half-struck marks, and inscriptions, these pieces can be very misleading.

On the other hand, consider these pieces when completely cleaned—and compare with true pieces completely cleaned—the differences then are narrowed. Michaelis feels that recast metal is of greater density—certainly some fakes have an indefinable smoothness. This touches on the difficulty of conveying in words the many almost imperceptible and very difficult-to-relate clues. Many very experienced collectors will tell you that they cannot

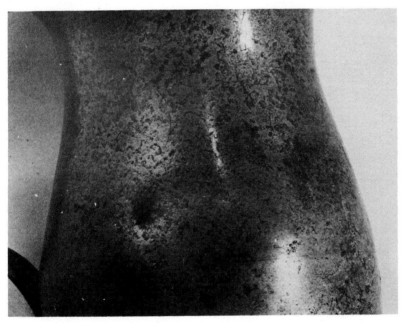

Fig. 97

Fake. The simulated corrosion is very realistic when first encountered, but its regularity and uniformity are soon remembered. This type first appeared in the 1920–30s, and a great many irresistible 'early' balusters, dishes, candlesticks, salts, and flagons were taken up unsuspectingly. Height to lip $5\frac{1}{4}''$.

(*The Pewter Society*)

177

put their finger on points, or describe them, yet show them a piece and their judgment will be instantaneous; even sometimes from a photograph. In cases of doubt, usually living with a doubtful piece for a few days, picking it up and putting it down, will bring the realisation and decision. So if you can, get it on approval!

The difficulty of describing features succinctly which appear in fakes will be appreciated, and also the reluctance to give too much away to any would-be faker, but some notes may be of help.

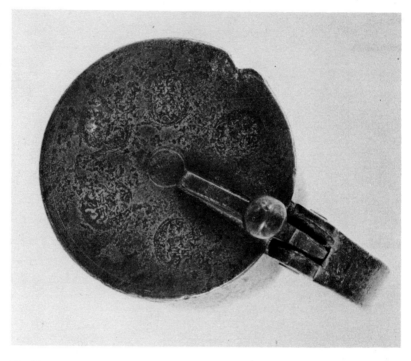

Fig. 98
Fake pieces, and true pieces too, had additional glamour pressed onto them as fake house marks. These are usually by dies far inferior to those actually used in the sixteenth and seventeenth centuries. In this example they are struck less firmly than is usual. This is another view of Fig. 97. There is no 'maker's' mark, but hR is distinctively struck on the rim. Lid $3\frac{3}{8}''$ diameter.

(*The Pewter Society*)

Fig. 99
Fake. The glamorous Horned Headdress is the spoon most frequently faked. The detail is quite good, except that the mould appears never to have been heated adequately, hence the metal solidified in the bowl before completely filling the mould. The stem should have been about $\frac{1}{4}''$ longer.

Most fakes carry an unnatural appearance of corrosion. They are speckled, too regular and too widespread. Compare a suspicious piece with known genuine pieces, in as near similar condition as possible, and the difference becomes obvious. We have mentioned the dull, solid, leaden, almost greasy impression sometimes given. Do not overlook the likelihood of different sources and different treatments. There may also be methods of accelerating the formation of oxide, but it is usually taken that any oxide which is not of the utmost hardness and brittleness is suspect. One may need to prod the scale in an inscription, or scratch the scale on the base. Sometimes the glamour is overloaded, such as adding interesting house marks to a baluster, to make it the more irresistible to induce even the suspicious to take a chance. Look for true scratched wear on bases, and the bruising of the thumbpiece on the handle—and that it 'fits'.

Very often there is some foolish error, such as the use of a spoon-maker's touch on a flagon or plate. I had a French broad-rim plate once, with a spurious English spoon-maker's touch on it. Spoons with badly cast bowls are invariably wrong, e.g. the

Fig. 100

A remarkably fine work of restoration by
an amateur—the lid and thumbpiece of
this flagon of c. 1625 were made in 1970,
and have been treated to tone perfectly
with the 340-year-old drum. If dusty and
in bad light, few people would spot that
any restoration has been carried out.
Body maker 'T.P.' Height to lip 9".

(*Mr S. Shemmell*)

frequent Horned Headdress fakes. Badly made dies for touch marks, and house marks too, almost certainly condemn a piece.

Sometimes desirable features are accentuated, such as over-big thumbpieces on dud hammerhead balusters purporting to be of c. 1650–c. 1700, as on that illustrated. The range of products which come to mind are primarily hammerhead balusters, then early flagons, broad-rim dishes, seventeenth-century salts and candlesticks, spoons, flat-lid tankards. A thought as to prices shows that awareness and discrimination can save hundreds of pounds.

The next 'fakes' to consider are 'made-up' pieces. By this I mean the baluster, tankard or flagon whose thumbpiece, let us say, was missing and has been replaced. A thumbpiece is not very intrusive—but what if it were a lid? A corroded body calls for an ancient-looking lid, so *some* stage of treatment is necessary. How far one accepts nitric-acid dulling, dents, scratches, induced warping, etc., can, I suppose only be decided by the buyer. The

Fig. 101
Reproduction pieces—not made to deceive—but some have acquired a little oxide, and often deceive the proprietor of a general antique shop—and at times, the panels of Antique Fairs.

Fig. 102
Some exhibits of the Pewter Society's
'Chamber of Horrors' at their Exhibition
at Reading, 1969. Some styles were
dreamed up, others are exactly true to
style. Some illustrations published pre-
war are very suspect.

ethics, and a code of practice, have not been established. Massé
decided early to stamp every single repair he carried out with his
initials. I would like to see every newly replaced part deeply but
unobtrusively stamped, but not all owners or repairers would
concur. Let us leave it by saying, not only do you judge a piece as a
whole, but every detachable part must be scrutinised. If one part
is found to be suspicious, or replaced, redouble the scrutiny of the
others.

There are probably only few fakes of eighteenth-century ware—
but there are a lot of reproductions. Fakes set out to deceive, but
reproductions merely to look old! Since a large quantity were
made before the Second World War, many have acquired some

oxide in addition to the chemical discoloration. Beware of Bush and Perkins 'hall marks'. Beware of Samuel Duncombe 'hall marks'—others as well. Many true pieces bear them—but many more are 'repro.'. Another is of a tilted half touch mark—the line where it purports to fade is usually strongly struck! Porringers with two spindly handles, small plates, and rectangular-based candlesticks and tappit hens all bearing the 'hall marks' were amongst other things made. Again, the price asked for a piece is no criterion. I have seen a very obvious repro. broad-rimmed plate priced at £95 in a shop—and not very convincing 'flat-lid style' repros. at midway prices.

To return to an earlier theme, let your knowledge grow by reading and handling as much as possible, and let your collection grow in step, both in number and value, with knowledge most certainly including something of fakes. You need to take bold decisions—let them be bold through knowledge, not rash chances. You may consider it worth while to own one or two duds for reference. In any case it is as well to get them out of circulation. The Pewter Society will welcome gifts of fakes for its 'Chamber of Horrors'.

Appendix I

Museum Displays and Collections (Medieval)

Birmingham, Art Gallery.
Birmingham, Weoley Castle (very small, fine medieval).
Dublin. National Museum of Ireland.
Edinburgh, National Museum of Antiquities of Scotland.
Edinburgh, Royal Scottish Museum.
Glasgow, Art Gallery and Museum.
London, Victoria and Albert Museum.
Truro Museum.
(Many other museums have a few scattered pieces of interesting pewter on display.)

Displays (Romano-British)

Bath; Bury St Edmunds; Cambridge (Archaeology and Ethnology); Devizes; Elveden (nr. Thetford); Ipswich; London—British Museum, Guildhall Museum, London Museum; Northampton; Norwich; Oxford; Reading; Taunton.

Collections in Museums, not normally on show, but which may be seen by arrangement

Cambridge, Fitzwilliam.
London, Guildhall.

Norwich, Strangers Hall.
Norwich, Bridewell.
(Many museums have some interesting pewter not on display.)

There is also a very fine collection at the Pewterers Hall, London (which is not a museum). It may be seen, by prior arrangement, if convenient. The Pewter Society's collection of fakes is also housed at the Pewterers Hall, and may be seen as above.

Appendix II

A Short Bibliography

Cotterell, Howard H. *Old Pewter, Its Makers and Marks,* Batsford, London, 1929 (Reprinted 1963).

Kerfoot, J. B. *American Pewter, Its Makers and Marks,* Bonanza, New York, 1924.

Britannia Metal, U.S.A.

Michaelis, R. F. *Antique Pewter of the British Isles,* Bell, London, 1955.

British Pewter, Ward Lock, London, 1969.

Chats on Old Pewter (revised from Massé).

Price, F. G. Hilton. *Old Base Metal Spoons,* Batsford, London, 1908.

Ullyett, Kenneth. *Pewter Collecting for Amateurs,* Frederick Muller, London, 1967.

Victoria and Albert Museum. *British Pewter,* H.M.S.O., No. 4, London, 1960.

Welch, Charles. *History of the Worshipful Company of Pewterers,* Blades, East & Blades, London, 1902.

Wood, L. Ingleby. *Scottish Pewterware and Pewterers,* Geo. A. Morton, Edinburgh, 1907.

Catalogue of the Exhibition of British Pewterware, Reading, 1969.
Sets of twenty photographs of above in situ. These two items
produced by the Pewter Society—Hon. Sec. C. A. Peal, c/o
John Gifford Ltd.

Most of the books above are out of print, but can no doubt be
obtained by a Public Library, if only for reference.

Some Articles in Antique Journals

Cotterell, H. H. 'Dating the Pewter Tankard', *Connoisseur,* April
1932.
'Scottish Pewter Measures', *Connoisseur,* May 1931.
Graeme, A. V. Sutherland. 'Pewter Church Flagons', *Connoisseur,*
June 1946.
'Pewter Spoons', *Connoisseur,* December 1947.
Michaelis, R. F. 'English Pewter Porringers', *Apollo,* July, August,
September, October 1949.
'Capacity Marks on Old Pewter Wine Measures', *Antique
Collector,* August 1954.
'Decoration on English Pewterware', *Antique Collector,* October
1963, February, August, December 1964.
Minchin, C. C. 'A Berkshire Pewter Collection', *Apollo,* April and
July 1946.
Peal, Christopher A. 'Pewter Salts, Candlesticks and Some Plates',
Apollo, May 1949.
'Notes on Pewter Baluster Measures, and their Capacities',
Apollo, June 1949.
'Notes on Pewter Flagons', *Apollo,* May 1950.
'Romano-British Plates and Dishes', *Proceedings of the Cambridge
Antiquarian Society,* Vol. LX, 1967.
'An English Pewter Collection', *Antiques* (U.S.A.), August 1969.
'Latten Spoons', *Connoisseur,* April and July 1970.

Appendix III

Glossary, Index, Colour Plates, Figures (major references only)

1605 type: The oldest type of church flagon likely to be seen.
Pl. 6; p. 93.

A.R.: Official stamp of capacity check used only in Queen Anne's
reign. Relates to both ale and spirits. Fig. 15; pp. 66, 115.

Acorn: (a) Flagon—mid-eighteenth-century York type, whose
body is acorn-shaped. Fig. 55; p. 121.
  (b) Spoon, the knop of which is in the form of an acorn. Mostly
fifteenth and sixteenth centuries. Fig. 28; pp. 87, 90.

Ale jug: Distinctive rustic type of the end of eighteenth century.
Fig. 65; p. 129.

Antimony: The metal alloying tin which produces Britannia metal.
p. 172.

Art Nouveau: Accepted as a style of art of around 1890–1905,
involving sweeping curves for flower stems, etc., of which
Beardsley was a leading exponent. Fig. 77; p. 146.

Ball: (*a*) Baluster measure thumbpiece: (1) Early English, sixteenth century—Figs. 26, 31; pp. 89, 113. (2) Later English, seventeenth century—Pl. 12; p. 129. (3) Early Scottish, seventeenth century —Pl. 15; p. 162. (4) Late Scottish, nineteenth century—Fig. 85; p. 160.

(*b*) Finial of eighteenth-century tankard handles. Figs. 67, 69; pp. 130, 135.

(*c*) Spoon knop: (1) Writhen—Pl. 5. (2) 'Golf', or 'Globe'— Pl. 5.

Baluster: (*a*) Lidded bulbous measure denoted by thumbpiece type. Pl. 12; Figs. 22, 26, 31, 50, 51, 52, 64, 85, 86, 87, 90; pp. 89, 96, 113, 114, 115, 128, 161, 162, 163, 168.

(*b*) Tear drop in stem of candlesticks, chalices, Fig. 62; p. 126.

(*c*) Spoon knop, similar to Seal. Pl. 15; Fig. 28; p. 90.

Beaker: Handleless mug, the only type of which likely to be met is nineteenth century. Fig. 74; p. 142.

Beefeater: Flagon, c. 1660–c. 1680, the lid of which has the squashed appearance of a Beefeater's hat. Pl. 7; p. 99.

Bell: (*a*) Candlestick—exceedingly rare Tudor type with bell-like base. Fig. 27; p. 89.

(*b*) Salt—exceedingly rare Tudor two- or three-part salt, light-house shaped. Fig. 37; pp. 89, 103.

Bellied, or bulbous, measure: The ubiquitous nineteenth-century pub measure for beer (and spirit in the smallest sizes). Figs 70, 71, 87; pp. 137, 138, 139, 163.

'Billie and Charlies': Nineteenth-century spurious base-metal figures, after the names of the two ingenious fakers.

Bleeding bowl: Porringer-like receptacle for blood-letting, with graduations for quantity taken. Fig. 60; p. 124.

Board of Trade numbers: After 1878 most districts adopted the Board's system of numbers, e.g. VR 135. pp. 69, 70–75.

Bouge: The wall round the well of a plate or dish. pp. 94, 122.

Bowl dish: Type in favour c. 1635–c. 1660. Fig. 29; p. 94.

Bristol measure: Similar to the squat conical copper measures. p. 140.

Britannia Metal: Pl. 16; Figs. 93, 94, 95, 96 ; pp. 4, 6, 15, 17, 19, 28, 31, 171 et seq.

Broad rim: Exceptionally wide proportion of rim to radius. c. 1640–c. 1665. Figs. 39, 41; p. 105.

Broken handle: Handle formed by two curves. Eighteenth and nineteenth centuries. Figs. 67, 69; p. 135.

Bud: Thumbpiece of the penultimate type of English baluster measure. Figs. 51, 52; pp. 114, 115, 116, 162.

Bumpy bottom: Plate or dish, sixteenth and early seventeenth centuries. Figs. 24, 25, 29, 30; pp. 87, 88, 94, 95.

Bun lid: Self-explanatory flagon of c. 1625–c. 1640. Pl. 6; p. 93.

Capstan: Salt of that shape of c. 1675–c. 1700. Fig. 37; p. 103.

Cartouche: Of pewter, a stamped 'label' for address or sales message. Around 1700, and later. p. 64.

Caster: For sugar, pepper, or sand. p. 143.

Cellar: Corruption of 'salière', a container for salt.

Chairback: Closed, or open—thumbpiece of flagons and tankards in eighteenth and early nineteenth centuries. Fig. 69.

Chalice: Catholic vessel, as in 'sepulchural chalice'. Figs. 79, 89; pp. 151, 154, 167.

Charger: 18″ and over.

Chip carving: Roman and Anglo-Saxon method of indenting lines for decoration. Fig. 20; p. 83.

Chopin: Scottish measure of capacity equal to half Scots pint, or $1\frac{1}{2}$ pints Imperial. Pl. 14; p. 157.

Chi Rho: Greek letters used by the Romans for Christ, being the first two letters. Fig. 18; p. 81.

Collar salt: Bearing a wide, usually octagonal, collar. Fig. 37; pp. 103, 104.

Communion cup: Fig. 79; pp. 151, 154.

Copper pewter: The better-grade medieval pewter used copper instead of lead as the adulterant. Used for plates and dishes. Fig. 24; pp. 26, 87.

Corrosion: The breakdown of the surface of metals by oxidisation. Fig. 3; pp. 24, 25.

'Cott. O.P.': H. H. Cotterell's *Old Pewter, Its Makers and Marks.* Still the standard reference, although very far from complete in the range of styles, and comprehensiveness of marks shown. pp. 17, 37, 41 et seq.

Crested: Scottish tappit hen with knop on the lid. Pl. 14.

Cup salt: Shaped like a grape-fruit cup. c. 1750–c. 1800. Fig. 61; p. 125.

Dean of the Guild: The head of a group of allied trades or crafts in Scotland. Fig. 85; pp. 149, 150.

Decoration: Cast—p. 95. Chip carving—Fig. 20; p. 83. Engraving —Pl. 11; Figs, 63, 64, 73; pp. 127, 141. Punched—Figs. 41, 44. Wriggled—Figs. 43, 45, 46, 73; pp. 108, 109, 141.

Denticulated projections: The small tooth-like projections on the front of lids of flat-lid tankards and flagons. Figs. 11, 49; p. 112.

Diamond: Knop of a spoon, like a solid arrow-head. Pl. 5; Fig. 28; pp. 87, 91.

Dish: 10″–18″ diameter.

Doll's-house pewter: Fig. 75; p. 143.

Dome lid: Lid styles which ran from c. 1680 to the nineteenth century on tankards and flagons. Pl. 9; Figs. 48, 49; p. 111.

Double measure: Two measures mounted opposing on the same stem, like a double-cupped chalice. One cup is twice the capacity of the other.

Double volute: The final type of English baluster, denoted by the similarity of the thumbpiece to the decoration on Greek columns. Fig. 64; p. 128.

Drip-tray: Like a ballet skirt round the stem of a candlestick to protect the hand from molten tallow. Fig. 38; p. 104.

Drum: Body of a tankard or flagon. Fig. 69; p. 134.

Edinburgh measure: Like an English bellied measure with a lid with concave curve. Fig. 87; p. 163.

Embryo shell: The last type of Scottish baluster measure, its thumb-piece most nearly resembling a smooth, ill-formed bivalve shell. Fig. 85; p. 160.

Entasis: Slightly convex curve of the 'straight' sides of a (column, originally) tankard, to counteract the hollow appearance of taper-ing straight lines. Pl. 9; p. 111.

Eruption: Result of internal oxidisation, causing surface bubbles. Fig. 5; p. 27.

Faculty: The necessary permission of a bishop for a church to sell any possessions. The interest only may be used, from the capital so raised.

Fake: Made to deceive. Figs. 97, 98, 99, 102; pp. 11, 176 et seq., et alia.

Fillet: Raised rib round the drum—particularly c. 1700–c. 1720. Fig. 69; p. 134.

Finial: (*a*) Knop of a spoon. Pl. 5; Figs, 28, 33; pp. 90, 97.
  (*b*) Terminal of the lower end of a handle. Fig. 69.
  (*c*) Knop on the lid of a flagon.

Flagon: Of a church—to carry wine to the table. Also of domestic use. Pls. 6, 7, 8, 10; Figs. 21, 54, 55, 63, 79, 80, 81, 88, 89; pp. 84, 93, 99, 120, 121, 127, 151 et seq., 164.

Flat lid: Distinctive lid style on tankards and flagons from c. 1650 to c. 1700. Pl. 8; Figs. 11, 47; p. 109.

Fosse Way: The Roman road through Lincoln to Ilchester, patrolling the north-west limit of the Civil Zone. p. 79.

Fothergill: Bristol maker who made a distinctive type of measure. p. 140.

Friendly Societies: Club within a local trade for mutual insurance, with considerable ceremonial. From them Trades Unions and Insurance developed. Fig. 63; p. 127.

GIV: Verification stamp of Imperial capacity used only in his reign. Fig. 15; pp. 66, 68.

Gadrooned: Raised cast beading, at a steep angle to the edge. Figs. 37, 38; pp. 103, 104.

Glasgow measure: Two styles of late Scottish measures with baluster bodies, but lids with single and double domes, respectively. Fig. 87; p. 163.

Guernsey measure: Type confined to Guernsey. Style strongly influenced by Normandy, but the makers were Channel Isle men who worked in London. Fig. 92; p. 169.

hR: Capacity check mark used in the latter half of the seventeenth century. So far the exact meaning has not been resolved, but may have been applied by the Clerk of the Markets under the control of the King's (or Queen's) Household (*h*ousehold *R*egis). Fig. 15; pp. 65, 66.

'Hall marks': In quotes, for they only simulate the true Hall marks of the Goldsmiths. On pewter they were of the makers' own designs, but frequently were exactly as the true silver marks. Used to make pewter as like silver as possible. Figs. 11, 91; pp. 60, 61, 62, 150, 168.

Hammerhead: Baluster measure denoted by the thumbpiece, which is like two outward-facing hammer faces. Often faked. Figs. 22, 31, 50; pp. 85, 96, 113.

Haystack: Irish measure, the shape being reminiscent of a conical old-fashioned haystack—or an oast house. Fig. 90; p. 167.

Hexagon: Spoon knop, like a six-sided crown. Fig. 28; p. 90.

Hollow-ware: Flagons, tankards, measures, etc.

Horned Headdress: Most attractive knop of a spoon, showing the contemporary headdress of c. 1485. Rare—and there are many fakes. Pl. 5; Fig. 99; pp. 87, 181.

House mark: Stamp on some measures and plates denoting establishment ownership, looking like a touch mark. Genuine ones are always from well-made irons. Figs. 13, 14, 56, 64, 98; pp. 64, 65, 122, 137, 179.

Jersey measure: Distinctive of Jersey. Text of 'Guernsey' is equally applicable. Fig. 92; p. 170.

Journeyman: Not a traveller or representative, but an employed workman who has been admitted to the Company, but not working on his own as a master. p. 40.

Knop: (a) The knob of a lid.
(b) The solid figure or head on the end of the handle of most medieval spoons.
(c) The ball-shaped knob on a candlestick stem. Fig. 38; p. 104.

L.T.P.: London Touch Plate (there are five). The number quoted in conjunction with L.T.P. is that of a mark's number in sequence. The original plates were lost in the Fire of London. p. 41.

Latten: Brass in sheet form, imported until 1565, when it was first made in this country. The spoon-makers were quick to see, and use, its advantages. pp. 86, 92.

Laver: Scottish vessel for baptismal water. Always with a spout, usually similar to but smaller than a flagon. Sometimes lidless. Fig. 80; pp. 152, 153.

Lias: Close-textured rock very well suited for casting the fine work in small Romano-British pewter products. p. 131.

Linear engraving: The letters are formed primarily by parallel lines of suitable lengths and changes of weight. Fig 73; p. 141.

Loggerheads: Capstan-shaped ink well.

Loving cup: For drinking together ceremonially, the cup being passed from one to another. p. 102.

Mace head: Many different designs of head mounted on staffs of about 2' long. In Britannia metal. Pl. 16.

Made-up: (a) Parts belonging elsewhere assembled to make one piece.
(b) Piece with new parts replacing lost ones, e.g. a lid. Fig. 100; p. 181.

Maidenhead: Attractive knop on sixteenth- and seventeenth-century spoons. Fig. 28; p. 90.

Mark: The impressed stamp of a maker's touch mark. Figs. 7, 8, 9, 10, etc.; pp. 36 et seq.

Melon: Segmented knop of fifteenth- and sixteenth-century spoons. Fig. 28; p.90.

Mutchkin: Scottish measure of $\frac{1}{4}$ Scots pint (15 fl. oz.). p. 157.

Narrow rim: Edge of $\frac{1}{2}''$–1'' wide, with comparatively heavy moulded and/or cast reedings. Latterly it developed a flat on which 'hall marks' were stamped. Fig. 42; pp. 105, 106.

Noggin: (a) Type of Irish measure. Fig. 90; p. 167.
(b) Capacity equal to one gill.

O.E.W.S.: Old English Wine Standard, which is five-sixths Imperial Measure. Usually only in English baluster measures. (Still the standard fluid measure of capacity of the U.S.A.) pp. 96, 128.

Paten: Plate for Communion bread. p. 154.

Pear measure; Late Scottish lidded measure, from Glasgow. Fig. 87; p. 163.

Pewter Collectors Club of America: p. 20.

Pewter Society: Figs. 97, 98, 102; pp. 8, 16, 20, 183; App. I, App. II.

Pip: Small blip of metal on the front of some Hammerhead, and Bud, thumbpieces.

Plate: 7″–10″.

Plouk: Filling mark in the larger sizes of tappit hen.

Pocks: Holes caused by pockets of corrosion, probably due to impurities in the alloy. Figs, 4, 6; pp. 26, 34.

Porringer: Pewter vessel for any thick liquid. Usually bowl-shaped, with, in England, one ear. Fig. 36; pp. 95, 102, 124.

Portcullis: The emblem for Imperial Measure, used in the verification stamp of 1826 to c. 1830. Fig. 16; p. 68.

Portrait spoon: Trifid with bust of William, William and Mary, or Anne. It is remarkable that the portraits are on spoons of pewter only. The space within the border is left blank on silver spoons. Fig. 35; p. 101.

Potbelly: One of the earliest types of Scottish measure known. c. 1680–c. 1740. Strong Continental influence. Fig. 83; p. 157.

Provenance: Whence it came—maker, owner, or locality.

Punched: Punching is almost always secondary to some other form of decoration used in the same piece. Figs. 41, 44.

Ramshorn: Thumbpiece with spiral points, almost only on flat-lid tankards. Figs. 47, 69; pp. 101, 135.

Reed: Moulding, usually cast, round the edge of late seventeenth-century plates and dishes; and single, in the early eighteenth-century. Figs. 41, 42, 43; pp. 105, 106, 107.

Reproduction: Not made to deceive—only to look like. The distinction from Fake is usually more apparent than this definition implies. Fig. 101; pp. 11, 20, 21, 146, 159, 182.

Restoration: Getting into a condition fairly similar to the original, including providing new parts where necessary, and making the appearance of the new in keeping with the old. Fig. 100; pp. 181, 182.

Romano-British: Made in Britain directly under Roman occupation influence. Figs. 18, 19; pp. 79, 80, 81.

Rose and Crown: (a) Originally a mark of quality, then of goods exported from England, and finally adopted by the makers as a secondary mark for prestige, applied close to their touch marks. Figs. 4, 6, 12; pp. 62, 63.
　(b) Very commonly used in Continental touches. Fig. 1; p. 4.
　(c) Used by inns of that name in house marks.

Saucer: 7″ and less.

Salt, A: Container for salt at table. Figs. 37, 61; pp. 102, 103, 125.

Scale: Hard oxide on pewter. Pl. 4; Fig. 3; pp. 24, 28, 29.

Scottish Measure: Scots pint—60 fl. oz. Chopin—30 fl. oz. Mutchkin—15 fl. oz. p. 157.

Scroll: Thumbpiece on eighteenth-century tankards and flagons faintly like a rolled scroll. Figs. 49, 69.

Seal: Knop on pewter and latten spoons. Two types, Simple and Gauntlet. Fig. 28; p. 90.

Secondary marks: Makers' subsidiary marks ('hall marks', his Rose and Crown, address labels, advertising slogans). Figs. 11, 12; pp. 60 et seq.

Shovel: Unusual late nineteenth-century thumbpiece on Scottish balusters. p. 161.

Single reed: Type of plate with one cast ring. c. 1700–c. 1735. Fig. 43; p. 107.

Slip Top: Sixteenth- and seventeenth-century spoon without knop. The end is usually cut at an angle, likened to a pruning cut. Figs. 28, 32, 33; pp. 27, 97.

Society flagon, tankard: For Friendly Society officials at meetings. Pl. 11; Fig. 63; pp. 126, 127.

Society of Pewter Collectors: p. 20.

Spire flagon: Tall dignified type of mid-eighteenth century. One of of the best styles ever made in pewter. Pl. 10; pp. 120, 121.

Spoon drawings: (a) Bowl, and stem, c. 1200–c. 1665. Fig. 23; pp. 85, 86.
   (b) Handles, c. 1660–c. 1710. Fig. 34; p. 99.

Spray: Thumbpiece on a type of baluster, Scottish or north of England, early eighteenth century. Pl. 15; p. 163.

Spun: Method of forming the drums of Britannia-metal ware, as in silver. p. 173.

Standish: Inkstand.

'Straight': Distinctive line in the handle design of Scottish (chiefly) measures. Fig. 82; p. 157.

Strike: To record a maker's touch on the touch-plate.

Stump end: Very rare type of fifteenth-century English spoon, with no knop. The stem tapers *towards* the bowl. Pl. 5.

Subsidiary mark: See *'Secondary mark'*.

Tankard: In this book taken as a drinking mug with or without lid. Pl. 9; Figs. 47, 67, 68, 69; pp. 109–13, 130–3, 141.

Tappit hen: (*a*) Style of Scottish measure. Pl. 14; Fig. 78; pp. 147, 157–9.
(*b*) Scottish pint size of above.

Tavern pot: Lidless drinking mug belonging to an inn. Pl. 9; p. 112.

Terminal date: When a pewterer ceased manufacture for any reason.

Thistle: Late Scottish type shaped like an opening thistle flower. All ordered to be destroyed, as they did not comply with the official angle of emptying, therefore very rare. Fig. 84; p. 159.

Tin pest: Decomposition of pure tin, caused erratically by low temperature. Pewter is stable and immune. pp. 25, 27, 28, 89.

Token, Communion: Issued to those found suitable by catechism to take Communion. Scottish, chiefly. Pl. 13; pp. 153, 154.

Touch: To touch is to impress a mark on a suitable surface.

Toy pewter: Fig. 75; p. 143.

Trade cards: Advertising business cards. Fig. 53; pp. 118, 119, 120.

Trencher salt: Low salt; always for use, not ceremonial. Figs. 37, 61; pp. 103, 125.

Triple grooved: Rare type of broad-rim dish with three deep grooves near the edge of the rim. c. 1660. Fig. 41; p. 105.

Triple reed: The most common item left of the seventeenth century —dish with three raised mouldings near the edge. Figs. 41, 42; pp. 105, 107.

Tulip: Tankard with low centre of gravity, and flared-out throat. c. 1720 to late nineteenth century. Figs. 67, 69; p. 134.

Twin cusp: Thumbpiece, chiefly on Beefeater flagons, comprised of two hemispheres. Pl. 7; p. 99.

VR: Verification stamp of Queen Victoria:
(*a*) with county or borough arms, c. 1837–1878. Fig. 17; p. 68.
(*b*) with district number, 1878–1902. p. 69.

Verification: Stamped proof of capacity check. Figs. 15, 16, 17; pp. 65, 66, 68, 69, 70–75, 164.

'Vickers White Metal': The first 'Britannia' metal, c. 1760. p. 172.

WIV: Verification mark of Imperial capacity used only in his reign. Fig. 15; pp. 66, 68.

WR: Verification of William III Act involving both ale, and wines and spirits. Used during his reign, apparently not during that of Anne, but then used through Georges I, II, and III up to 1826. Fig. 15; pp. 65, 66, 128, 129.

Wavy edge: Type of heavily reeded edge on plates and dishes for a short while c. 1760, adopted from France. Fig. 57; p. 122.

Wedge: Baluster thumbpiece type: probably not a type of complete thumbpiece—merely showing loss of a Ball, or Hammerhead. Therefore probably not really a type. Figs. 13, 14, 31; p. 96.

West Country: Measure in conical form, similar to those in copper. Fig. 72; p. 140.

Wood, L. Ingleby: Wrote the only book to date, on Scottish pewter. Now known not to be very comprehensive, but is nevertheless very instructive. p. 148.

Worshipful Company of Pewterers: One of the old City Companies, who safeguarded their guild. They now own a very fine collection. In the text referred to as the Pewterers Company. pp. 8, 15, 36–40, 62, 63, 92.

Wriggling: Attractive English decoration, the lines of which are formed 'waddling', zig-zag, a flat-ended tool, pivoting on the edges. Chiefly c. 1670–c. 1710. Figs. 43, 45, 46, 73; pp. 108, 109, 141.

X: Originally a mark of quality. May stand for 10:1 proportion of tin:lead. Or may be short for 'Extraordinary ware called Hard mettle'. pp. 60, 63, 64.

York flagon: (a) Straight-sided, c. 1730. Fig. 55; p. 121.
(b) Acorn-shaped, c. 1745. Fig. 55; p. 121.